**Leckie**

the education publisher
**for Scotland**

# 1st Level Maths
## Assessment Pack

**Authors:** Karen Hart and Michelle Mackay
**Assessment Consultant:** Carol Lyon
**Series Editor:** Craig Lowther

001/17022021

10 9 8 7 6 5 4 3 2 1

ISBN 9780008392475

*Published by*
Leckie & Leckie Ltd
An imprint of HarperCollinsPublishers
Westerhill Road, Bishopbriggs, Glasgow, G64 2QT
T: 0844 576 8126 F: 0844 576 8131
leckieandleckie@harpercollins.co.uk
www.leckieandleckie.co.uk

HarperCollins Publishers
1st Floor, Watermarque Building,
Ringsend Road,
Dublin 4,
Ireland

Publisher: Fiona McGlade
Project manager: Rachel Allegro

*Special thanks to*
Copy editor: Mitch Fitton
Layout and illustration: Jouve
Proofreaders: Louise Robb, Deborah Dobson and Lauren Reid

A CIP Catalogue record for this book is available from the British Library

*Acknowledgements*
Images © Shutterstock

# Contents

Find all the downloadable resources – including Yearly Progress Checks, End of Level Assessments and record sheets – here: collins.co.uk/primarymathsforscotland

# Introduction

This Assessment Pack includes high quality assessments for Numeracy and Mathematics. It can be used to support ongoing professional judgements of children's progress and next steps within the Scottish Curriculum for Excellence (CfE) First Level. Assessment Packs for Early Level and Second Levels are also available, to ensure consistency from Preschool to P7.

As stated in *Building the Curriculum 5: A framework for assessment*, we at Leckie understand that ongoing assessment and pupil feedback is vital to, '…support learning and promote learner engagement resulting in greater breadth and depth in learning, including a greater focus on the secure development of knowledge, understanding and skills', Scottish Government (2011).

In this introduction, you will find:

1.  Guidance for assessment

2.  Introduction to the assessments supplied in this pack

3.  Support for record keeping

4.  Tables showing coverage of the Experiences and Outcomes and Benchmarks

## Guidance for assessment

Ongoing formative assessment is an essential part of the learning process. Black and Wiliam (2009) listed five key principles of learning:

1.  Clarifying and sharing learning intentions and success criteria;

2.  Engineering effective classroom discussions and other learning tasks that elicit evidence of student understanding;

3.  Providing feedback that moves learners forward;

4.  Activating students as instructional resources for one another; and

5.  Activating students as the owners of their own learning.

The Leckie Primary Maths for Scotland Textbooks and Teacher Guides were designed and written with these principles at the forefront, to support teachers to plan and deliver excellent mathematical learning. Both resources create opportunities for: diagnostic questions that assess prior learning; observation of and feedback on day-to-day learning (including outdoor learning); leading and listening to learning conversations.

The Leckie Primary Maths for Scotland Assessment Packs now add to this by providing rich, diagnostic tasks and activities to help the educator track and monitor progress and achievement. The materials contained within the packs are both summative and formative in nature. Not only do they provide important information about the child's current knowledge and skills (summative information), they also help diagnose misconceptions and identify next steps for both the educator and child (formative information).

The Leckie Teacher Guides contain advice and support that will help the educator address common misconceptions that children experience as they learn and develop mathematical skills, thus helping to ensure smooth and continuous progress.

Throughout the learning, teaching and assessment process, feedback should be used to ensure the pupil can answer these questions:

*   Where am I going?
*   How am I going there?
*   Where to next?

Hattie (2012)

# Primary Maths for Scotland First Level Assessment Pack

The Primary Maths for Scotland First Level Assessment Pack contains two types of assessment:

1. Yearly Progress Checks

2. End of Level Assessments

# Yearly Progress Checks (YPC)

These assessments provide full coverage of the CfE First Level Experiences and Outcomes. Each Progress Check assesses mental agility, procedural fluency and conceptual understanding of the key numeracy and mathematical ideas appropriate to this level.

## When to use the Yearly Progress Checks

There are three Yearly Progress Checks included in this pack. Yearly Progress Check 1A is designed to be used at the end of Textbook 1A. Yearly Progress Check 1B is designed to be used at the end of Textbook 1B. Yearly Progress Check 1C is designed to be used at the end of First Level which, for almost all children, will be at the end of P4.

## How to use the Yearly Progress Checks

Yearly Progress Checks can be completed in groups or by the whole class. The questions have been arranged in the order of the Experiences and Outcomes for ease of tracking, however teachers should use their own professional judgement in deciding how the assessment is administered and bundled together.

The Yearly Progress Checks are a blend of questions where the child will need to verbalise their answer and explain their thinking, and questions where the child will show you their understanding using manipulatives and written answers. There are photocopiable resources for ease of use.

The Yearly Progress Checks include a mix of bare number problems, questions using pictures and diagrams and contextualised problems. Throughout the assessment, the pupil is encouraged to explain their thinking to help you to understand the strategy they are using. This is important as the same question can be solved in many, increasingly sophisticated ways.

At times, the pupil will become stuck on a question; this is to be expected. It may be appropriate for you to read or restate the question, perhaps in a context with which the pupil is more familiar. Be mindful, however, not to make the question easier or to give away a strategy that the pupil can use. Wherever a pupil struggles, this is evidence of next steps in learning.

The answers the pupil provides will be an indication of the mathematical and numerical facts that they have committed to memory and the strategies and skills they are using to solve problems.

Misconceptions will also be highlighted by the responses given. The Primary Maths for Scotland Teacher Guides list possible misconceptions by topic and provide help and suggested tasks for the pupil to undertake. By comparing the data collected from a group or class it will also be possible to check how well a particular topic was learned and whether the issue lies with the individual or the group.

## Using the Marking Guidance

The Marking Guidance provided for the Yearly Progress Checks includes a reference to the appropriate Experiences and Outcomes linked to each question. The 'Notes' section explains what to expect of a child who is 'on track' within the level and provides advice on 'further learning required' for the child who has struggled or could not answer the question. Where further learning is required, the Leckie Teacher Guide describes common misconceptions and offers advice on appropriate learning activities to help address these.

# Introduction

Question:

5. a  Draw a line on the sandwich to split it in half.

Marking Guidance:

| Topic – benchmarks / Es & Os | Question | Answer | Notes |
|---|---|---|---|
| **Fractions, decimal fractions and percentages**<br>MNU 1-07a<br>MNU 1-07b<br>MNU 1-07c | 5a | Sandwich cut into two equal parts | **On track**<br>• Appreciates that to halve the sandwich means to divide it into two equal parts and that there is more than one way to do this.<br><br>**Further learning required**<br>• May divide the sandwich into more than two parts.<br>• May divide the sandwich into two unequal parts. |

# End of Level Assessments (EOL)

These assessments are contextualised, holistic assessments that blend together topics from across the Mathematics and Numeracy Curriculum. There are several questions per assessment that can be used to provide a broad picture of the pupil's mastery of the national standards set out in the Benchmarks.

The questions will be in new and unfamiliar contexts to the pupil. This will increase the challenge for them as they have to both understand the context and work out what strategy, skills and knowledge to use. As with the Yearly Progress Checks, the pupil is encouraged to show their thinking to help you understand the strategies they are using and identify any misconceptions.

## When to use the End of Level Assessments

As the name suggests, the End of Level Assessments should be used when you judge that a child or group of children has mastered the First Level Mathematics Curriculum, or to ascertain whether they have achieved mastery in instances where you are unsure. This will generally be at the end of P4 if the child is meeting age-related expectations.

## How to use the End of Level Assessments

The End of Level Assessments are a blend of activities and questions linked to a context. It is advisable that the children are introduced to the context before they are assessed on their Numeracy and Mathematical skills and knowledge. This can be done through storytelling, playing games, discussion and linking the assessment activity to previous learning.

The assessments can be completed in groups or by individuals. When children complete the assessment tasks in groups, they will discuss their ideas and allow the teacher to observe the skills and knowledge they have mastered. Throughout the assessment it is permitted for the children to use concrete materials, drawings, diagrams, etc., if necessary.

## Using the Marking Guidance

The Marking Guidance for the End of Level Assessments includes the Experiences and Outcomes for each question part, the question, a description of the answer that tells you the pupil is 'on track' and a 'review' section that describes what you can expect a child to do if they have not mastered the concept assessed. The Leckie Teacher Guide has further guidance on how to support pupils who were unable to engage in the assessment tasks, to help move their learning forwards.

| Assessment | Topic | Question |
|---|---|---|
| Q5b | Number and number processes<br><br>MNU 1-02a | **On track**<br>9 sweets. Child may say 90 divided by 10 is 9, explain that they counted from 873 in tens and knew to stop at 963 because they had counted ten 9 times (perhaps 'double-counting' on their fingers), or may say they counted how many numbers they had written down, equating one number with 1 sweet.<br><br>**Review**<br>Encourage child to act out or visualise the problem. |

## Record Keeping

### Yearly Progress Checks Record Keeping

There are record sheets online (collins.co.uk/primarymathsforscotland) for you to use to track the progress of the pupils against each question. We suggest that you 'mark' each child as they complete their First Level Yearly Progress Check. We suggest the ABC coding described below but this can easily be replaced with green/amber/red to provide a quick visual check:

A.  Chose an appropriate strategy/method and used it correctly (green)

B.  Chose an appropriate strategy/method but used it incorrectly or made an error in calculation (amber)

C.  Chose an inappropriate strategy/method or did not attempt the question (red)

### End of Level Assessment Record Keeping

We have provided record sheets for you to use as the children complete the End of Level Assessments and suggest using the coding:

- O. On track
- R. Review

Again, a simple colour coding system can be used to provide a quick visual check, e.g. using green for 'on track' and red for 'review'.

It is important to note that a child does not need to successfully complete every task contained within the Leckie Primary Maths for Scotland Assessment Pack to achieve CfE First Level. This decision will be based on the teacher's professional judgement as they observe their pupils engaged in their day-to-day learning. Nevertheless, we hope the record sheets included will provide a simple reference to where children have mastery of key Numeracy and Mathematics concepts and where they require further support.

## References

Black P. and Wiliam D. (2009) Developing the theory of formative assessment. *Educational Assessment, Evaluation and Accountability,* 21(1), 5–31

Hattie, J.A.C. (2012) *Visible Learning for Teachers.* London. Routledge

Scottish Government (2011) *Curriculum for Excellence. Building the Curriculum 5: A framework for assessment.* Available online https://www.education.gov.scot/Documents/btc5-framework.pdf [accessed on 07/10/2020]

# Introduction

## Tables showing coverage of the CfE Numeracy and Mathematics Experiences and Outcomes and Benchmarks

| Curriculum organisers | Experiences and outcomes | Benchmarks to support practitioners' professional judgement of achievement of a level | Coverage in First Level Assessment Pack |
|---|---|---|---|
| Estimation and rounding | I can share ideas with others to develop ways of estimating the answer to a calculation or problem, work out the actual answer, then check my solution by comparing it with the estimate.<br><br>MNU 1-01a | • Uses strategies to estimate an answer to a calculation or problem; for example, doubling and rounding.<br><br>• Rounds whole numbers to the nearest 10 and 100 and uses this routinely to estimate and check the reasonableness of a solution. | YPC A, B, C<br>EOL 1 |
| Number and number processes | I have investigated how whole numbers are constructed, can understand the importance of zero within the system and can use my knowledge to explain the link between a digit, its place and its value.<br><br>MNU 1-02a<br><br>I can use addition, subtraction, multiplication and division when solving problems, making best use of the mental strategies and written skills I have developed.<br><br>MNU 1-03a | • Reads, writes, orders and recites whole numbers to 1000, starting from any number in the sequence.<br><br>• Demonstrates understanding of zero as a placeholder in whole numbers to 1000.<br><br>• Uses correct mathematical vocabulary when discussing the four operations including, subtract, add, sum of, total, multiply, product, divide and shared equally.<br><br>• Identifies the value of each digit in a whole number with three digits; for example, $867 = 800 + 60 + 7$.<br><br>• Counts forwards and backwards in 2s, 5s, 10s and 100s.<br><br>• Demonstrates understanding of the commutative law; for example, $6 + 3 = 3 + 6$ or $2 \times 4 = 4 \times 2$.<br><br>• Applies strategies to determine multiplication facts; for example, repeated addition, grouping, arrays and multiplication facts.<br><br>• Solves addition and subtraction problems with three-digit whole numbers.<br><br>• Adds and subtracts multiples of 10 or 100 to or from any whole number to 1000.<br><br>• Applies strategies to determine division facts; for example, repeated subtraction, equal groups, sharing equally, arrays and multiplication facts. | YPC A, B, C<br>EOL 1, 4<br><br><br><br><br><br><br><br><br><br>YPC A, B, C<br>EOL 2, 3, 4, 5 |

# Introduction

| Curriculum organisers | Experiences and outcomes | Benchmarks to support practitioners' professional judgement of achievement of a level | Coverage in First Level Assessment Pack |
|---|---|---|---|
| | | • Uses multiplication and division facts to solve problems within the number range 0 to 1000. | |
| | | • Multiplies and divides whole numbers by 10 and 100 (whole number answers only). | |
| | | • Applies knowledge of inverse operations (addition and subtraction; multiplication and division). | |
| | | • Solves two-step problems. | |
| Fractions, decimal fractions and percentages | Having explored fractions by taking part in practical activities, I can show my understanding of:<br><br>• how a single item can be shared equally<br>• the notation and vocabulary associated with fractions<br>• where simple fractions lie on the number line.<br><br>MNU 1-07a<br><br>Through exploring how groups of items can be shared equally, I can find a fraction of an amount by applying my knowledge of division.<br><br>MNU 1-07b<br><br>Through taking part in practical activities including use of pictorial representations, I can demonstrate my understanding of simple fractions that are equivalent.<br><br>MTH 1-07c | • Explains what a fraction is using concrete materials, pictorial representations and appropriate mathematical vocabulary.<br><br>• Demonstrates understanding that the greater the number of equal parts, the smaller the size of each share.<br><br>• Uses the correct notation for common fractions to tenths; for example, $\frac{1}{2}$, $\frac{2}{3}$ and $\frac{5}{8}$.<br><br>• Compares the size of fractions and places simple fractions in order on a number line.<br><br>• Uses pictorial representations and other models to demonstrate understanding of simple equivalent fractions; for example, $\frac{1}{2} = \frac{2}{4} = \frac{3}{6}$.<br><br>• Explains the role of the numerator and denominator.<br><br>• Uses known multiplication and division facts and other strategies to find unit fractions of whole numbers; for example, $\frac{1}{2}$ or $\frac{1}{4}$. | YPC A, B, C<br>EOL 2, 6<br><br><br><br><br><br><br><br><br><br><br><br>YPC A, B, C<br>EOL 4, 5<br><br><br><br><br>YPC C |

# Introduction

| Curriculum organisers | Experiences and outcomes | Benchmarks to support practitioners' professional judgement of achievement of a level | Coverage in First Level Assessment Pack |
|---|---|---|---|
| Money | I can use money to pay for items and can work out how much change I should receive.<br><br>MNU 1-09a<br><br>I have investigated how different combinations of coins and notes can be used to pay for goods or be given in change.<br><br>MNU 1-09b | • Identifies and uses all coins and notes to £20 and explores different ways of making the same total.<br>• Records amounts accurately in different ways using the correct notation; for example, 149p = £1.49 and 7p = £0.07.<br>• Uses a variety of coin and note combinations to pay for items and give change within £10.<br>• Applies mental agility number skills to calculate the total spent in a shopping situation and is able to calculate change.<br>• Demonstrates awareness of how goods can be paid for using cards and digital technology. | YPC A, B, C<br>EOL 2, 5<br><br><br><br><br><br>YPC A, B, C<br>EOL 2, 5 |
| Time | I can tell the time using 12-hour clocks, realising there is a link with 24-hour notation, explain how it impacts on my daily routine and ensure that I am organised and ready for events throughout my day.<br><br>MNU 1-10a<br><br>I can use a calendar to plan and be organised for key events for myself and my class throughout the year.<br><br>MNU 1-10b<br><br>I have begun to develop a sense of how long tasks take by measuring the time taken to complete a range of activities using a variety of timers.<br><br>MNU 1-10c | • Tells the time using half past, quarter past and quarter to using analogue and digital 12-hour clocks.<br>• Records 12-hour times using a.m. and p.m. and is able to identify 24-hour notation; for example, on a mobile phone or computer.<br>• Records the date in a variety of ways, using words and numbers.<br>• Uses and interprets a variety of calendars and 12-hour timetables to plan key events.<br>• Knows the number of seconds in a minute, minutes in an hour, hours in a day, days in each month, weeks and days in a year.<br>• Orders the months of the year and relates these to the appropriate seasons.<br>• Selects and uses appropriate timers for specific purposes. | YPC A, B, C<br>EOL 2, 5<br><br><br><br><br><br><br><br><br><br>YPC A, B, C<br>EOL 5, 6<br><br><br><br>YPC A, C |

# Introduction

| Curriculum organisers | Experiences and outcomes | Benchmarks to support practitioners' professional judgement of achievement of a level | Coverage in First Level Assessment Pack |
|---|---|---|---|
| Measurement | I can estimate how long or heavy an object is, or what amount it holds, using everyday things as a guide, then measure or weigh it using appropriate instruments and units.<br><br>MNU 1-11a<br><br>I can estimate the area of a shape by counting squares or other methods.<br><br>MNU 1-11b | • Uses knowledge of everyday objects to provide reasonable estimates of length, height, mass and capacity.<br>• Makes accurate use of a range of instruments including rulers, metre sticks, digital scales and measuring jugs when measuring lengths, heights, mass and capacities using the most appropriate instrument for the task.<br>• Records measurements of length, height, mass and capacity to the nearest standard unit; for example, millimetres (mm), centimetres (cm), grams (g), kilograms (kg), millilitres (ml), litres (l).<br>• Compares measures with estimates.<br>• Uses knowledge of relationships between units of measure to make simple conversions; for example, 1 m 58 cm = 158 cm.<br>• Reads a variety of scales on measuring devices including those with simple fractions; for example, $\frac{1}{2}$ litre.<br>• Uses square grids to estimate then measure the areas of a variety of simple 2D shapes to the nearest half square.<br>• Creates shapes with a given area to the nearest half square using square tiles or grids.<br>• Recognises that different shapes can have the same area (conservation of area). | YPC A, B, C<br>EOL 1, 4, 5<br><br><br><br><br><br><br><br>YPC A, C |
| Mathematics – its impact on the world, past, present and future | I have discussed the important part that numbers play in the world and explored a variety of systems that have been used by civilisations throughout history to record numbers.<br><br>MTH 1-12a | • Investigates and shares understanding of the importance of numbers in learning, life and work.<br>• Investigates and shares understanding of a variety of number systems used throughout history. | YPC A, B, C<br>EOL 1 |

# Introduction

| Curriculum organisers | Experiences and outcomes | Benchmarks to support practitioners' professional judgement of achievement of a level | Coverage in First Level Assessment Pack |
|---|---|---|---|
| Patterns and relationships | I can continue and devise more involved repeating patterns or designs, using a variety of media.<br><br>MTH 1-13a<br><br>Through exploring number patterns, I can recognise and continue simple number sequences and can explain the rule I have applied.<br><br>MTH 1-13b | • Counts forwards and backwards in 2s, 5s and 10s from any whole number up to 1000.<br>• Describes patterns in number; for example, in the multiplication tables and hundred square.<br>• Continues and creates repeating patterns involving shapes, pictures and symbols.<br>• Describes, continues and creates number patterns using addition, subtraction, doubling, halving, counting in jumps (skip counting) and known multiples. | YPC A, B<br><br><br><br><br><br>YPC A, B, C<br>EOL 1 |
| Expressions and equations | I can compare, describe and show number relationships, using appropriate vocabulary and the symbols for equals, not equal to, less than and greater than.<br><br>MTH 1-15a<br><br>When a picture or symbol is used to replace a number in a number statement, I can find its value using my knowledge of number facts and explain my thinking to others.<br><br>MTH 1-15b | • Understands and accurately uses the terms 'equal to', 'not equal to', 'less than', 'greater than', and the related symbols $(=, \neq, <, >)$ when comparing quantities.<br>• Applies understanding of the equals sign as a balance, and knowledge of number facts, to solve simple algebraic problems where a picture or symbol is used to represent a number; for example, $\blacklozenge + 17 = 30$ and $\blacklozenge \times 6 = 30$. | YPC B<br>EOL 4<br><br><br><br><br><br>YPC A, B, C<br>EOL 6 |

# Introduction

| Curriculum organisers | Experiences and outcomes | Benchmarks to support practitioners' professional judgement of achievement of a level | Coverage in First Level Assessment Pack |
|---|---|---|---|
| Properties of 2D shapes and 3D objects | I have explored simple 3D objects and 2D shapes and can identify, name and describe their features using appropriate vocabulary.<br><br>MTH 1-16a<br><br>I can explore and discuss how and why different shapes fit together and create a tiling pattern with them.<br><br>MTH 1-16b | • Names, identifies and classifies a range of simple 2D shapes and 3D objects and recognises these shapes in different orientations and sizes.<br><br>• Uses mathematical language to describe the properties of a range of common 2D shapes and 3D objects including side, face, edge, vertex, base and angle.<br><br>• Identifies 2D shapes within 3D objects and recognises 3D objects from 2D drawings.<br><br>• Identifies examples of tiling in the environment and applies knowledge of the features of 2D shapes to create tiling patterns incorporating two different shapes. | YPC A, B, C<br>EOL 5<br><br><br><br><br><br><br><br>YPC B<br>EOL 3 |
| Angle, symmetry and transformation | I can describe, follow and record routes and journeys using signs, words and angles associated with direction and turning.<br><br>MTH 1-17a<br><br>I have developed an awareness of where grid reference systems are used in everyday contexts and can use them to locate and describe position.<br><br>MTH 1-18a<br><br>I have explored symmetry in my own and the wider environment and can create and recognise symmetrical pictures, patterns and shapes.<br><br>MTH 1-19a | • Uses technology and other methods to describe, follow and record directions using words associated with angles, directions and turns, including full turn, half turn, quarter turn, clockwise, anticlockwise, right turn, left turn, right angle.<br>• Knows that a right angle is 90°.<br>• Knows and uses the compass points, north, south, east and west.<br>• Uses informal methods to estimate, compare and describe the size of angles in relation to a right angle.<br>• Finds right angles in the environment and in well-known 2D shapes.<br>• Identifies where and why grid references are used.<br>• Describes, plots and uses accurate two-figure grid references, demonstrating knowledge of the horizontal and vertical location.<br>• Identifies symmetry in patterns, pictures, nature and 2D shapes.<br>• Creates symmetrical pictures and designs with more than one line of symmetry. | YPC A, B, C<br>EOL 2, 3<br><br><br><br><br><br><br><br>YPC B, C<br>EOL 5<br><br><br><br><br><br>YPC A, B, C<br>EOL 5 |

# Introduction

| Curriculum organisers | Experiences and outcomes | Benchmarks to support practitioners' professional judgement of achievement of a level | Coverage in First Level Assessment Pack |
|---|---|---|---|
| Data and analysis | I have explored a variety of ways in which data is presented and can ask and answer questions about the information it contains.<br><br>MNU 1-20a<br><br>I have used a range of ways to collect information and can sort it in a logical, organised and imaginative way using my own and others' criteria.<br><br>MNU 1-20b<br><br>Using technology and other methods, I can display data simply, clearly and accurately by creating tables, charts and diagrams, using simple labelling and scale.<br><br>MTH 1-21a | • Asks and answers questions to extract key information from a variety of data sets including charts, diagrams, bar graphs and tables.<br><br>• Selects and uses the most appropriate way to gather and sort data for a given purpose; for example, a survey, questionnaire or group tallies.<br><br>• Uses a variety of different methods, including the use of digital technologies, to display data; for example, as block graphs, bar graphs, tables, Carroll diagrams and Venn diagrams.<br><br>• Includes a suitable title, simple labelling on both axes and an appropriate scale where one unit represents more than one data value in graphs. | YPC A, B, C<br>EOL 2, 6<br><br><br><br><br>YPC A, C<br>EOL 4<br><br><br><br><br>YPC B, C<br>EOL 4 |
| Ideas of chance and uncertainty | I can use appropriate vocabulary to describe the likelihood of events occurring, using the knowledge and experiences of myself and others to guide me.<br><br>MNU 1-22a | • Uses mathematical vocabulary appropriately to describe the likelihood of events occurring in everyday situations, including probable, likely/unlikely, certain/uncertain, possible/impossible and fair/unfair.<br><br>• Interprets data gathered through everyday experiences to make reasonable predictions of the likelihood of an event occurring. | YPC A, B, C<br>EOL 4, 6 |

# Yearly progress check 1A

## Resources

**Resources needed and available in the setting**

Ruler · Squared paper

## Questions

1. a Draw a ring around the set that you think has more bears. Count to check.

b Is 8 closer to 0 or 10?

You may use the number line to help you.

_____

c Estimate how many cars there are. Group the cars to check your answer.

_____

d  How many cars are there?

2.  a  Write the numbers the teacher tells you.  _____

b  Match each number to its name.

| | |
|---|---|
| 11 | nineteen |
| 12 | fifteen |
| 13 | sixteen |
| 14 | eighteen |
| 15 | twenty |
| 16 | fourteen |
| 17 | twelve |
| 18 | eleven |
| 19 | seventeen |
| 20 | thirteen |

c  How many straws altogether?

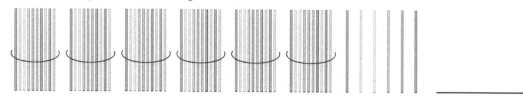

_____

d  How many groups of ten can you see?

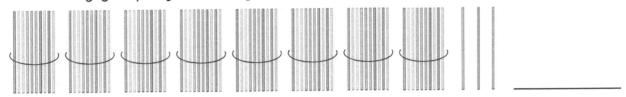

_____

e  Fill in the missing numbers.

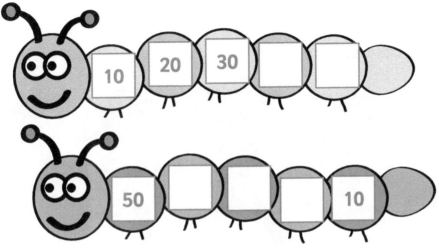

f  Isla thinks **57** says *fifty-seven*.   Amman thinks it says *seventy-five*.

Who is correct?  _____

Why?  _____

g  Write 87 in words.  _____

Write the number that comes:

h  Before 50.  _____

i  After 37.  _____

j  Before 81.  _____

k  After 99.  _____

l  Inbetween 29 and 31.  _____

m  Nuria is counting on in 10s. Write the missing numbers.

7, 17, ☐ , ☐ , ☐ , ☐

n   Amman is counting back in tens. Write the missing numbers.

[  ] , [  ] , [  ] , [  ] , 74, [  ]

o   I would like 15 straws, how many more do I need?

_____

10 + [  ] = 15

p   Isla is watching a running race.

| Maggie | Lola | Logan | Olivia | Lucy |

Who is first?   _____

Who is third?   _____

Who is fifth?   _____

3.  a   Write two addition facts for this domino.

_____      _____

b   Write a fact family for this ten frame.

_____      _____

c   Double 6 =   _____

d   Double [  ] = 18

e   16 = double [  ]

f   Isla has 10 dolls and 2 teddies. How many toys is this?

_____

g   Amman has some cars and trucks. There are 15 toys altogether. 10 are cars.
How many trucks are there?

_____

h   9 + 5 = ☐

i   7 + 6 = ☐

j   Nuria is on number 32. She jumps forward 6. What number is she on now?

Use the number line to help you.

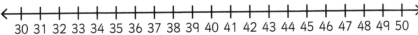

_____

k   Finlay is on number 58. He jumps forward 4. What number is he on now?

Use the number line to help you.

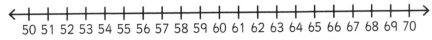

_____

l   Finlay was on number 40. He jumps back 7. What number is he on now?
Use the number line to help you.

←┼┼┼┼┼┼┼┼┼┼┼┼┼┼┼┼┼┼┼┼→
20 21 22 23 24 25 26 27 28 29 30 31 32 33 34 35 36 37 38 39 40

_____

m   If 7 − 4 = 3, what is 17 − 4?   _____

n   Find the missing number and write down the number sentence.

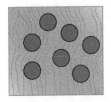   and   *some* counters under here   **makes 11 altogether**

7        +        ?              = 11

_____

o  Find the missing number and write down the number sentence.

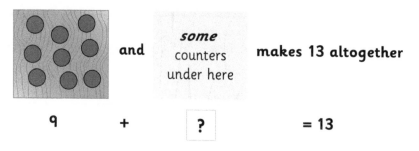

9       +       ?           = 13

---

p  There were 14 cats altogether. 6 were in the garden and the rest were in the house.

How many were in the house? Write a number sentence to show your thinking.

q  Lucy had 9 hair clips. She found 8 more in her bag. How many does she have now? Write a number sentence to show your thinking.

# Yearly progress check 1A

r   There were 18 cows in a field. 9 went to be milked.

How many cows are in the field now? Write a number sentence to show your thinking.

s   Finlay had some football stickers. Isla gave him 11 more. Now he has 16.

How many football stickers did Finlay have at the start? Write a number sentence to show your thinking.

t   Nuria has 8 dolls and 5 teddy bears. How many more dolls than teddy bears are there? Write a number sentence to show your thinking.

4.  a   Share 14 sweets equally between two plates.

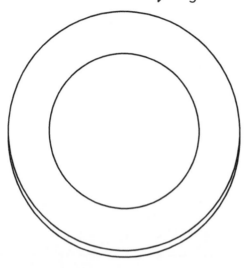 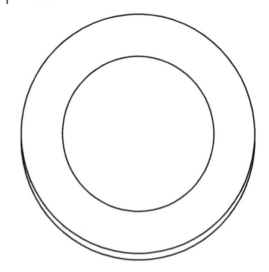

How many sweets on each plate?  _____

b   Isla has 10 sandwiches. She puts 2 on each plate. How many plates does she need?

_____

c

i   How many rows?  _____

ii  How many columns?  _____

iii How many dots altogether?  _____

d   How many dots can you see? How did you work it out?

_____

e   Nuria is skip counting forward in twos. Write the missing numbers.

2, 4,  , 8,  , 12

f   Amman is skip counting backwards in twos. Write the missing numbers.

[    ] , 16, 14, [    ] , 10

g   Isla is skip counting forwards in tens. Write the missing numbers.

20, 30, [    ] , 50, [    ] , [    ]

h   Amman is skip counting backwards in tens. Write the missing numbers.

90, [    ] , 70, [    ] , [    ] , 40

i   How many fingers and thumbs altogether?

_____

j   Amman is having a party. He has 25 packets of crisps. He wants to put five packets on each table. How many tables will he need?

k   Make an array to show this problem: six groups of two.

Can you write a number sentence to show this? Can you skip count to get the total?

l  Isla has 18 stickers. She shares them equally between 6 friends.

How many stickers will each friend get?

m  Finlay has 20 pencils. He puts five in each tub. How many tubs does he need?

n  Nuria has 16 eggs and puts them into 2 boxes.

How many eggs are in each box?

5. a  Draw a line on the sandwich to split it in half.

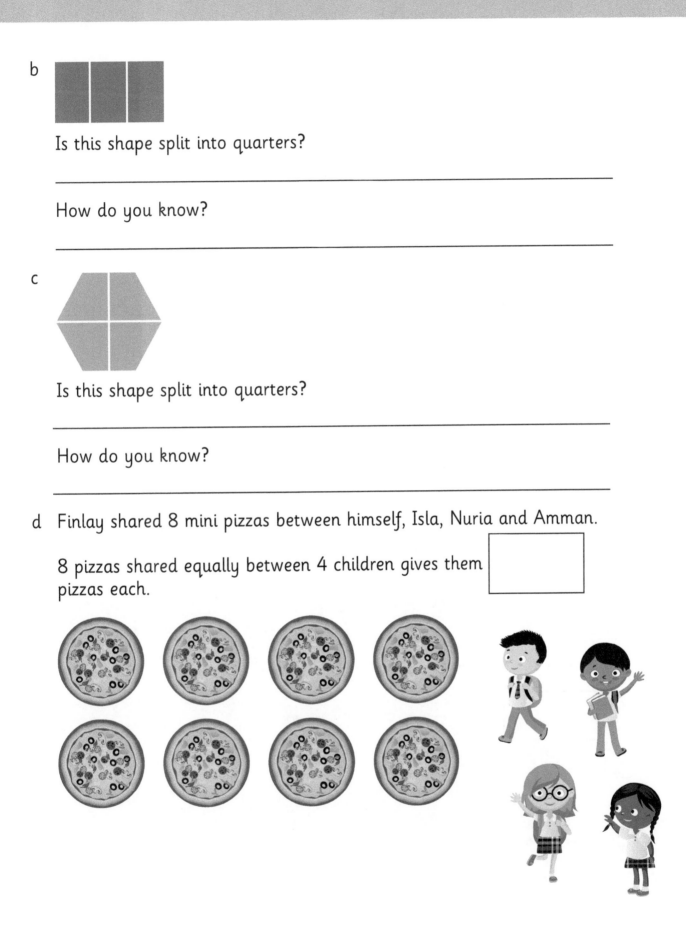

b

Is this shape split into quarters?

_____

How do you know?

_____

c

Is this shape split into quarters?

_____

How do you know?

_____

d Finlay shared 8 mini pizzas between himself, Isla, Nuria and Amman.

8 pizzas shared equally between 4 children gives them
pizzas each.

e   Nuria has 2 pizzas.

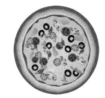

If she splits each pizza into quarters, how many pieces will she have?

f   Find one half of 12 sweets.

g   Find one quarter of 12 sandwiches.

h  Finlay is making boxes of cookies for his friends. Each box will have 3 cookies. He has 12 cookies in total.

How many boxes can he make?

i  Mrs Jones has 16 sweets. She wants to share them equally between Nuria, Finlay, Amman and Isla.

How many sweets will each person get?

6. a These toys are for sale at the school fair. Draw coins to show two different
   ways to pay for each toy.

b How much does Isla pay for food? How much does she pay for toys?

6p        8p              10p        4p        6p

_____          _____

c Amman has a 20p coin. How much **change** would he get if he spent:

13p

15p

2p

10p

d   Find each total. Draw the **least number of coins** needed to pay each total.

i   11p + 9p =

Coins:

ii  8p + 6p =

Coins:

iii 3p + 2p + 2p =

Coins:

iv 8p + 8p =

Coins:

7.  a   What day is it today?   _____

b   What day was it yesterday?   _____

c   Here is a timetable for Primary 2.

|              | MONDAY   | TUESDAY  | WEDNESDAY | THURSDAY     | FRIDAY   |
|--------------|----------|----------|-----------|--------------|----------|
| 9.00–10.30   | Literacy | Numeracy | Literacy  | Numeracy     | Literacy |
| 10.30–10.45  | BREAK    | BREAK    | BREAK     | BREAK        | BREAK    |
| 10.45–12.15  | Numeracy | Literacy | Numeracy  | Literacy     | Numeracy |
| 12.15–1.00   | LUNCH    | LUNCH    | LUNCH     | LUNCH        | LUNCH    |
| 1.00–2.00    | Art      | P.E.     | Free Play | P.E.         | RME      |
| 2.00–3.00    | RME      | Assembly | Science   | Outdoor Play | Science  |

What day and time is free play?   _____   _____

Finlay loves outdoor play! What day is outdoor play on?

_____

How long does the class have Art for every week?

_____

d Draw a line to match each activity with the best unit of time.

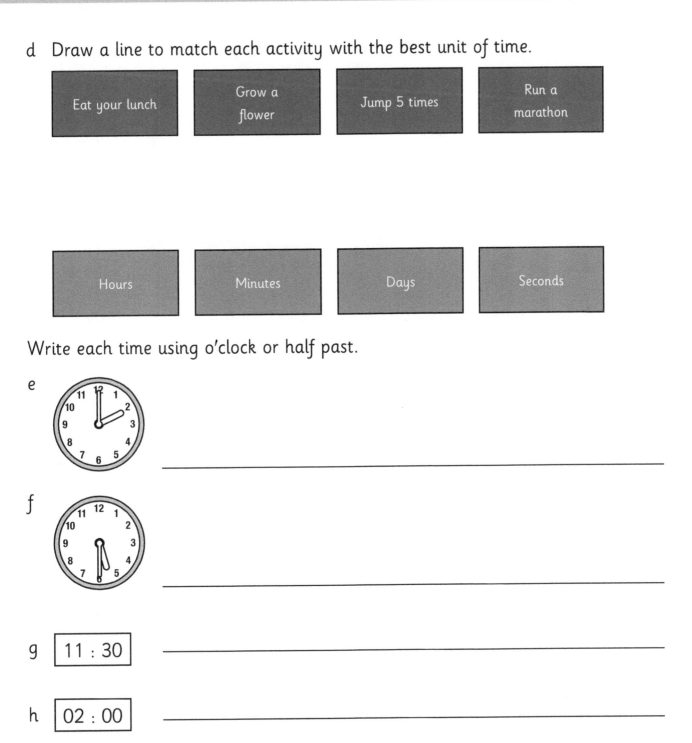

| Eat your lunch | Grow a flower | Jump 5 times | Run a marathon |

| Hours | Minutes | Days | Seconds |

Write each time using o'clock or half past.

e _____

f _____

g  11 : 30  _____

h  02 : 00  _____

8.  a

tower

car

mum    giraffe

Finlay

pencil

Which is tallest?  _____

Which is shortest?  _____

Order the objects from shortest to tallest.

_____

b

book    pencil

Which is lighter, the book or the pencil?  _____

How do you know?  _____

c

box    car

Which is heavier, the box or the car?  _____

How do you know?  _____

d

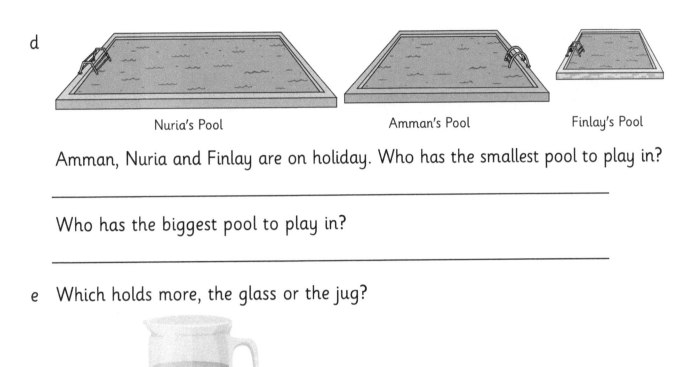

Nuria's Pool         Amman's Pool         Finlay's Pool

Amman, Nuria and Finlay are on holiday. Who has the smallest pool to play in?

_____

Who has the biggest pool to play in?

_____

e   Which holds more, the glass or the jug?

_____

f   Tick the bin that holds less.

g   About how many blocks tall is the teddy?

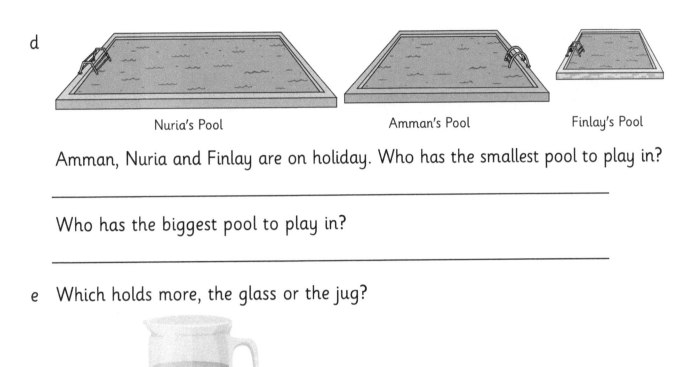

⬜ ⟵────── one block

The teddy is about ⬜ blocks tall.

h   About how many blocks long is the train?

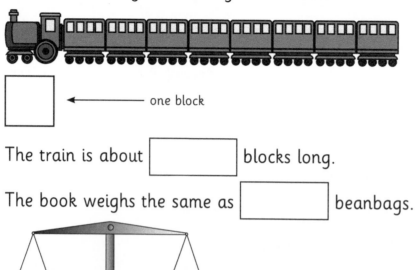

   ← one block

The train is about [          ] blocks long.

i   The book weighs the same as [          ] beanbags.

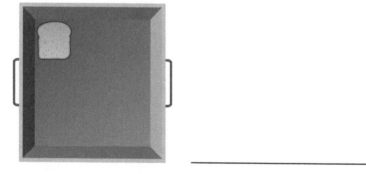

j   Estimate how many slices of toast will fit on the tray.

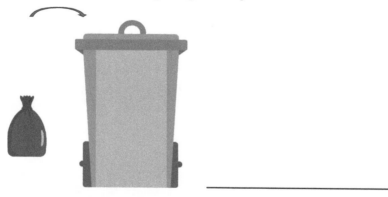

_____

k   Estimate how many bags will fit in the bin.

_____

9.  List three places outdoors where you might see numbers.

• 

• 

•

10. Draw the next two shapes in each pattern.

a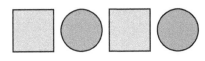

b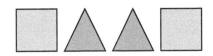

Write the next two numbers in each pattern.

c  6, 8, 10, 12, ☐ , ☐

d  3, 7, 11, 15, ☐ , ☐

11. Write a number or a symbol (+ or −) in the box to make each number sentence true.

a  ☐ + 10 = 15

b  14 = 6 + ☐

c  16 ☐ 10 = 6

d  7 ☐ 8 = 15

12. a  Draw a square and a rectangle.

b  Draw a triangle.

c  Draw a circle.

d   Look at these shapes.

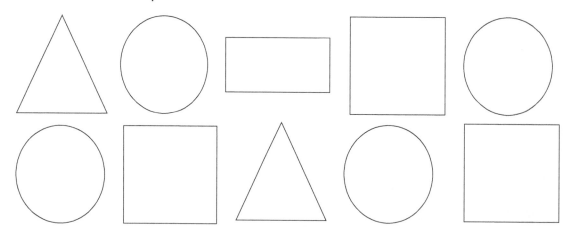

How many shapes have no corners?  _____

How many shapes have three corners?  _____

How many shapes have four corners?  _____

e   Look at these objects.

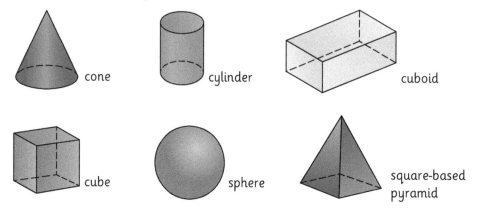

cone

cylinder

cuboid

cube

sphere

square-based
pyramid

i   Which objects have 6 faces?  _____

ii  Which objects have at least one curved face?  _____

13.  a

What is on the right of the teddy?  _____

What is on the left of the pot of clay?  _____

What is on the right of the football?  _____

What is below the rocket?  _____

What is above the tablet?  _____

What is in-between the rocket and the tablet?  _____

b   This is a plan of Jollytown.

Start at the police station [image]. Go down one square and left 2 squares.

You are at the _____.

Start at the library [image]. Go up one square and right 1 square. You are

at the _____.

Start at the fire station [image]. Go right 2 squares, down 2 squares,

left 2 squares. You are at the _____.

Complete each pattern to make it symmetrical.

c   [pattern squares]

d   [pattern squares]

14. Here are the results of a survey about favourite snacks.

   a   Complete the table: to show how many of each snack there is

| Snack | Tally | Total |
|---|---|---|
| Bag of crisps | | |
| Chocolate bar | | |
| Apple | | |
| Banana | | |
| Breadsticks | | |

   b   The most popular snack is _____.

c

Our favourite ice-cream    **Key:** = 1 person's favourite

Which flavour of ice cream is the more popular? _____

How many people like strawberry? _____

How many fewer people like lemon than strawberry? _____

d          **Finlay's family**

| brothers | 𝟀 | 𝟀 | 𝟀 | 𝟀 | | | | |
| sisters | 𝟀 | 𝟀 | | | | | | |

**Key:** 𝟀 = 1 person

Does Finlay have more brothers or sisters? _____

15. Answer these questions by saying if they are **certain, possible** or **impossible** to happen.

It will be sunny tomorrow. _____

I will stay the same age forever. _____

My dog will have to go to the vet. _____

| Topic – benchmarks / Es & Os | Question | Answer | Notes |
|---|---|---|---|
| **Estimation and rounding** MNU 1-01a | 1a | Circles the collection on the left, then counts each set to compare them. | **On track**<br>• Predicts which collection has more and checks by counting, systematically crossing off each bear in turn.<br>**Further learning required**<br>• May not understand the meaning of the term 'bigger'. |
| | 1b | 10 | **On track**<br>• Responds correctly by using the number line provided, or visualising the number line in their mind's eye.<br>**Further learning required**<br>• May not understand the meaning of the word 'closer' nor know the number sequence from 0 to 10. |
| | 1c | 14 (Accept a reasonable alternative estimate.) | **On track**<br>• Predicts how many cars are in the set then checks by grouping them into twos, fives or tens.<br>**Further learning required**<br>• May not understand that 'group' means form collections with the same number of cars in each.<br>• May not understand the term 'estimate'. |
| | 1d | 24 | **On track**<br>• Works out the total by subitising (10, 10 and 4) and counting in tens and ones.<br>**Further learning required**<br>• May not appreciate they must change from tens to ones when counting 24 objects, i.e. may say 10, 20, 30, 40, 50, 60 not 10, 20, 21, 22, 23, 24. |
| **Number – order and place value** MNU 1-02a | 2a | Correct number as per teacher's instructions. (The number given should be a 2-digit number greater than 20.) | **On track**<br>• Correctly writes number read out by teacher in numerals.<br>**Further learning required**<br>• May be unable to write the given number or does so incorrectly, perhaps reversing the digits. For example, writes twenty-five as 52. |
| | 2b | 11 – eleven<br>12 – twelve<br>13 – thirteen<br>14 – fourteen<br>15 – fifteen<br>16 – sixteen<br>17 – seventeen<br>18 – eighteen<br>19 – nineteen<br>20 – twenty | **On track**<br>• Correctly matches each number word to the correct numeral.<br>**Further learning required**<br>• May not read teen numbers as teens but rather 'ty', e.g. reads fifteen as fifty. |

| Topic – benchmarks / Es & Os | Question | Answer | Notes |
|---|---|---|---|
| | 2c | 66 | **On track**<br>• Counts in tens and ones to find the correct total.<br>**Further learning required**<br>• May not count each bundle of straws as ten, but ones, i.e. counts 1, 2, 3, 4, 5, 6, 7, 8, 9, 10, 11, 12.<br>• May not change from tens to ones when counting the collection, i.e. counts 10, 20, 30, 40, 50, 60, 70, 80 90, 100, 110, 120 (or stops after 100 or says 101, 102). |
| | 2d | 8 groups of ten | **On track**<br>• Identifies 8 groups of ten only, i.e. appreciates that the 3 individual straws each have a value of one.<br>**Further learning required**<br>• May count the individual straws as tens so answers 11 groups.<br>• May count the three individual straws as an extra group so answers 9. |
| | 2e | 40, 50<br><br>40, 30, 20 | **On track**<br>• Recognises that the pattern increases in values of 10 (first caterpillar) and decreases in values of 10 (second caterpillar).<br>**Further learning required**<br>• May only be able to count forwards and backwards in 1s.<br>• May attempt to count in tens but be unsure of the sequence, missing out or reversing the order of some numbers.<br>• May be able to count forwards in tens but find counting backwards more challenging. |
| | 2f | Isla is correct<br><br>5 in the tens column = 50; 7 in the ones column = 7<br>50 + 7 = 57 | **On track**<br>• Appreciates that 57 = 50 + 7.<br>**Further learning required**<br>• May 'reverse' 2-digit numbers when trying to read them, i.e. reads fifty-seven as seventy-five. |
| | 2g | eighty-seven | **On track**<br>• Writes 87 in words, making a reasonable attempt to spell the words eighty and seven correctly.<br>**Further learning required**<br>• May recognise the numeral 87 but have difficulty connecting this to the written word. |

| Topic – benchmarks / Es & Os | Question | Answer | Notes |
|---|---|---|---|
| | 2h | 49 | **On track**<br>• Response shows understanding of the terms 'number before' and 'in-between' and confidence in crossing a decade number (questions h and i). |
| | 2i | 38 | **On track**<br>• Response shows understanding of the term 'number after' and knowledge of the repeating 1-9 sequence between each pair of decade numbers (question i). |
| | 2j | 80 | **On track**<br>• Response shows understanding of the term 'number before' and knowledge of the repeating 1-9 sequence between each pair of decade numbers (question j). |
| | 2k | 100 | **On track**<br>• Response shows understanding of the term 'number after' and the need to cross into the 'hundreds' numbers after 99 (question k). |
| | 2l | 30 | **Further learning required**<br>• May not understand the language of before, after or in between.<br>• May struggle to cross the decades when counting forwards and/or backwards.<br>• May not have recall of the number sequence and will need to recount the number sequence from an earlier starting point, e.g. says 27, 28, 29, <u>30.</u> |
| | 2m | 27, 37, 47, 57 | **On track**<br>• Counts forwards and backwards in tens off the decade number to complete each sequence correctly. |
| | 2n | 34, 44, 54, 64, 84 | **Further learning required**<br>• May be unable to count in tens off the decade and counts in ones instead. This may result in a miscount.<br>• May be able to count forwards in tens, off the decade, but have difficulty with backward number word sequences of this type. |
| | 2o | 5 | **On track**<br>• Counts on from 10 to find the missing addend.<br>**Further learning required**<br>• May start the count at 1 instead of 10.<br>• May count on incorrectly, e.g. '10, 11, 12, 13, 14' and give the answer 14. |

| Topic – benchmarks / Es & Os | Question | Answer | Notes |
|---|---|---|---|
| | 2p | 1st Maggie<br>3rd Logan<br>5th Lucy | **On track**<br>• Correctly associates the words first, third and fifth with the positions 1, 3 and 5.<br><br>**Further learning required**<br>• May not understand ordinal numbers. |
| **Number – addition and subtraction**<br>MNU 1-03a | 3a | $6 + 4 = 10$<br>$4 + 6 = 10$ | **On track**<br>• Correctly writes two addition facts showing a commutative understanding.<br><br>**Further learning required**<br>• May not understand the commutative property and therefore only write one addition.<br>• May confuse addition and subtraction and write two subtraction facts. |
| | 3b | $4 + 6 = 10$<br>$6 + 4 = 10$<br>$10 - 6 = 4$<br>$10 - 4 = 6$ | **On track**<br>• Response shows understanding of fact families and has two additions and two subtractions made from the same whole and parts.<br><br>**Further learning required**<br>• Does not know what a 'fact family' is or how many facts are in a family.<br>• May be unable to connect the image to appropriate number sentences.<br>• May be able to identify two addition facts but find subtraction facts more challenging. |
| | 3c<br>3d<br>3e | 12<br>9<br>8 | **On track**<br>• Understands the word 'double' and responds correctly by:<br>  - adding two lots of six or by recalling a known fact, i.e. $6 + 6$ (question c).<br>  - thinking 'double/two lots of what makes 18' then sharing or grouping concrete materials or using a known fact, i.e. $9 + 9$ (question d).<br>  - thinking '16 is double what', making the link with halving then sharing concrete materials into two equal groups or using a known fact, i.e. $8 + 8$ (question e).<br><br>**Further learning required**<br>• May not understand that double means two lots of the same amount.<br>• May be able to represent and solve the missing sum example (question c) but have difficulty with missing addend examples (questions d and e). |

| Topic – benchmarks / Es & Os | Question | Answer | Notes |
|---|---|---|---|
| | 3f<br>3g | 12 toys<br>5 trucks | **On track**<br>• Choses an appropriate method (for example, using concrete materials, drawing a picture, number line or bar model, using known facts) to solve each problem.<br>• May write a number sentence that mirrors their actions/thoughts, i.e. to reflect how they solved the problem.<br><br>**Further learning required**<br>• Unable to select the correct operation, e.g. adds 10 and 15 in question g.<br>• May use the chosen method incorrectly, e.g. uses a number line incorrectly in question f and arrives at an answer of 11. |
| | 3h<br>3i | 14<br>13 | **On track**<br>• Chooses an appropriate strategy (for example counters and ten frames, 10 plus facts, doubles) to calculate accurately.<br><br>**Further learning required**<br>• May be unable to select an appropriate strategy or uses the chosen strategy incorrectly. |
| | 3j | 38 | **On track**<br>• Uses a graduated number line correctly to count on.<br><br>**Further learning required**<br>• May not trust counting on as a strategy.<br>• May not understand how to use a graduated number line and may count numerals instead of jumps. |
| | 3k<br>3l | 62<br>33 | **On track**<br>• Uses a graduated number line to correctly count on where they are required to cross a decade number.<br>• Uses a graduated number line to correctly count back from a decade number.<br><br>**Further learning required**<br>• May not trust counting on or back as a reliable strategy.<br>• May not understand how to use a graduated number line and may count numerals instead of jumps. |

| Topic – benchmarks / Es & Os | Question | Answer | Notes |
|---|---|---|---|
| | 3m | 13 | **On track**<br>• Can use a known fact (7 – 4) to generate a new fact (17 – 4).<br><br>**Further learning required**<br>• Does not understand the relationship between the two calculations and needs to subtract 4 items from 17 items or count back 4 from 17. |
| | 3n<br><br>3o | 4<br>7 + **4** = 11<br><br>4<br>9 + **4** = 13 | **On track**<br>There are many ways to solve these problems, below indicates various 'On track' strategies:<br>• Thinks 'part–part–whole' and represents the problem as a bar model.<br>• Counts on using a number line.<br>• Uses known addition and subtraction bonds.<br>• Uses concrete materials or drawings.<br><br>**Further learning required**<br>• Unable to select the correct operation, e.g. adds 7 + 11 and 9 + 13.<br>• May use the chosen method incorrectly, e.g. may count numerals instead of jumps on a number line. |
| | 3p<br><br><br><br>3q<br><br><br>3r<br><br><br>3s<br><br><br>3t | 8 cats<br>Child writes 6 + ? = 14 or<br>14 – 6 = ?<br><br>17 hair clips<br>Child writes 9 + 8 = 17 OR 8 + 9 = 17<br><br>9 cows<br>Child writes 18 – 9 = 9<br><br>5 stickers<br>Child writes ? + 11 = 16 or 16 – 11 = ?<br><br>3 teddy bears<br>Child writes 5 + ? = 8 or 8 – ? = 5 | **On track**<br>• Can visualise and/or represent each problem using concrete materials, drawings or diagrams and select an appropriate strategy to solve it, e.g.<br>  – joining and separating collections of objects<br>  – counting on or back on a number line<br>  – using known addition and subtraction bonds<br>  – thinking of the problem in terms of 'part–part–whole' and representing it as a bar model.<br>• Can write a number sentence that mirrors their thoughts and actions (see answer column).<br><br>**Further learning required**<br>• May be able to represent and solve standard, result unknown problems (i.e. questions Q and R) but find start and change unknown problems more challenging.<br>• May be able to represent and solve addition and subtraction problems where the action of adding some or taking some away is 'obvious' (questions P–S) but find questions that involve comparing two static amounts (question T) more challenging. |

| Topic – benchmarks / Es & Os | Question | Answer | Notes |
|---|---|---|---|
| | | | • May use the chosen method incorrectly, e.g. may count numerals instead of jumps on a number line. |
| | | | • May be unable to connect how they solved the problem to an appropriate number sentence, e.g. answers '8 cats' in question P but is unable to write either of the number sentences given or writes the number sentence incorrectly, e.g. $6 + 14 = 8$. |
| **Number – multiplication and division** MNU 1-03a | 4a | Child draws 14 sweets / dots and shares them equally between the two plates<br><br>7 on each plate | **On track**<br>• Correctly shares the sweets to make two equal groups of 7.<br><br>**Further learning required**<br>• May not know what 'equal' means.<br>• May 'share' the 'sweets' between the plates but make unequal groups, e.g. creates groups of 10 and 4; 8 and 6, etc. |
| | 4b | 5 plates | **On track**<br>• Successfully models the problem with concrete materials (grouping the 'sandwiches' in twos) and can identify the number of groups (plates).<br><br>**Further learning required**<br>• May confuse groups and shares and give the answer '2'. |
| | 4c | i 3 rows<br>ii 6 columns<br>iii 18 dots altogether | **On track**<br>• Correctly identifies the number of columns and rows in the array and skip counts or counts by ones to find the total number of dots.<br><br>**Further learning required**<br>• May not understand the words 'row' and 'column' or confuse the two.<br>• May be unfamiliar with arrays. |
| | 4d | 20 dots<br>Explanation of how they arrived at the total proves ability to subitise the dice pattern and skip count in fives to find a total. | **On track**<br>• Can count in fives to find the total number of dots.<br><br>**Further learning required**<br>• May be unable to skip count in fives and so resorts to counting the dots one at a time.<br>• May not understand the words row and column and confuse the two. |

# Yearly progress check 1A    Marking guidance

| Topic – benchmarks / Es & Os | Question | Answer | Notes |
|---|---|---|---|
| | 4e<br>4f | 6,10<br>18, 12 | **On track**<br>• Completes each sequence correctly by counting forwards and backwards in 2s and trusts that this will give the same result as counting in 1s.<br>**Further learning required**<br>• May be unable to identify the pattern, e.g. writes the numbers 5 and 9 because they come after 4 and 8.<br>• May lack confidence in reciting the number word sequence in twos.<br>• May find backwards number word sequences more challenging than forwards number word sequences. |
| | 4g<br>4h | 40, 60, 70<br>80, 60, 50 | **On track**<br>• Completes each sequence correctly by counting forwards and backwards in 10s and trusts that this will give the same result as counting in 1s.<br>**Further learning required**<br>• May be unable to identify the pattern.<br>• May lack confidence in reciting the number word sequence in tens.<br>• May find backwards number word sequences more challenging than forwards number word sequences. |
| | 4i | 30 fingers | **On track**<br>• Skip counts the fingers in fives and trusts that this will give the same result as counting in 1s.<br>**Further learning required**<br>• May not trust skip counting as a strategy, preferring to count in ones. |
| | 4j | 5 tables | **On track**<br>• May represent and solve the problem, by drawing or forming equal groups with concrete materials, and understands that each group represents one table.<br>• May represent and solve the problem by 'double counting', i.e. counts in fives, possibly using fingers to keep track of the count, knows to stop at 25 and appreciates that each finger represents one table.<br>• May be able to visualise the problem as 5 groups of 5 and uses repeated addition (5 + 5 + 5 + 5 + 5) or a known fact (5 × 5) to solve it. |

| Topic – benchmarks / Es & Os | Question | Answer | Notes |
|---|---|---|---|
| | | | **Further learning required** <br> • May be unable to represent and solve the problem in any of the ways described above. <br> • May answer '5' without fully understanding the problem, i.e. shares the 'crisps' between 5 tables and thinks the answer is five because there are five packets of crisps on each. *Presenting the child with a similar problem using different numbers will help determine their ability to define between grouping and sharing tasks, e.g. 10 packets of crisps, how many tables? |
| | 4k | .. <br> .. <br> .. <br> .. <br> .. <br> .. <br><br> 2 + 2 + 2 + 2 + 2 + 2 = 12 <br> OR 2 x 6 = 12 <br> OR 6 x 2 = 12 | **On track** <br> • Successfully uses rows and columns to draw an array. <br> • Successfully uses skip counting in 2s to get the correct answer. <br> • Correctly writes a number sentence related to the array. <br> **Further learning required** <br> • May not understand what an array is. <br> • May not understand what column and row mean. <br> • May not be able to skip count and can only count in 1s. <br> • May be unable to write a number sentence to match the array. |
| | 4l | 3 stickers | **On track** <br> • Understands what it means to 'share equally'. <br> • Uses concrete materials, drawings or known facts to solve the problem. <br> **Further learning required** <br> • May not understand the concept of equal shares, e.g. may 'share' 18 items between 6 people, giving some to each, but be unconcerned about whether or not they each get the same amount. |
| | 4m | 4 | **On track** <br> • May represent and solve the problem by drawing or forming equal groups with concrete materials, and understands that each group represents one tub. <br> • May represent and solve the problem by 'double counting', i.e. counts in fives, possibly using fingers to keep track of the count, knows to stop at 20 and appreciates that each finger represents one tub. |

| Topic – benchmarks / Es & Os | Question | Answer | Notes |
|---|---|---|---|
| | | | • May be able to visualise the problem as 4 groups of 5 and uses repeated addition (5+5+5+5) or a known fact (4 × 5) to solve it. |
| | | | **Further learning required** |
| | | | • May be unable to represent and solve the problem in any of the ways described above. |
| | | | • May answer '4' without fully understanding the problem, i.e. shares the 'pencils' between 5 tubs and answers four because there are four pencils in each. |
| | 4n | 8 | **On track** |
| | | | • Understands what it means to 'share equally'. |
| | | | • Uses concrete materials, drawings or known facts to solve the problem. |
| | | | **Further learning required** |
| | | | • May not understand the concept of equal shares, e.g. may 'share' 16 items between 2 boxes, giving some to each, but be unconcerned about whether or not they each get the same amount. |
| **Fractions, decimal fractions and percentages** MNU 1-07a MNU 1-07b MNU 1-07c | 5a | Sandwich cut into two equal parts | **On track** |
| | | | • Appreciates that to halve the sandwich means to divide it into two equal parts and that there is more than one way to do this. |
| | | | **Further learning required** |
| | | | • May divide the sandwich into more than two parts. |
| | | | • May divide the sandwich into two unequal parts. |
| | 5b | No, because the shape is split into 3 equal parts | **On track** |
| | | | • Appreciates that to split something into quarters you divide it into 4 equal parts. |
| | | | **Further learning required** |
| | | | • May not understand what quarters are or what they look like, i.e. four equal parts. |
| | | | • May answer 'no' but be unable to justify their choice, e.g. by explaining there are only three equal parts. |

| Topic – benchmarks / Es & Os | Question | Answer | Notes |
|---|---|---|---|
| | 5c | Yes, because the shape is split into 4 equal parts | **On track**<br>• Appreciates that to split something into quarters you divide it into 4 equal parts.<br><br>**Further learning required**<br>• May not understand what quarters are or what they look like, i.e. four equal parts.<br>• May answer 'yes' but be unable to justify their choice, e.g. by explaining there are four equal parts. |
| | 5d | 2 pizzas each | **On track**<br>• Understands what it means to 'share equally'.<br>• Uses concrete materials, drawings or known facts to solve the problem.<br><br>**Further learning required**<br>• May not understand the concept of equal shares, e.g. may 'share' the pizzas, giving some to each, but be unconcerned about whether or not they each get the same amount. |
| | 5e | 8 pieces (quarters) | **On track**<br>• Splits each pizza into four equal parts.<br>• Understands that quarters mean four equal parts and that two whole pizzas will provide eight quarters.<br><br>**Further learning required**<br>• May not understand what quarters are or what they look like.<br>• May be able to quarter the pizzas by splitting each into four equal parts but believe there cannot be more than four quarters altogether. |
| | 5f | 6 sweets | **On track**<br>• Shares 12 items equally into two equal groups and identifies how many in each group.<br><br>**Further learning required**<br>• May be able to equate 'one half' with 'one of two equal parts' but have difficulty understanding that one half of a quantity can be made up of more than one item. |
| | 5g | 3 sandwiches | **On track**<br>• Shares 12 items equally into four equal groups and identifies how many in each group.<br><br>**Further learning required**<br>• May be able to equate 'one quarter' with 'one of four equal parts' but have difficulty understanding that one quarter of a quantity can be made up of more than one item. |

| Topic – benchmarks / Es & Os | Question | Answer | Notes |
|---|---|---|---|
| | 5h | 4 boxes | **On track**<br>• Groups 12 items into threes and can identify how many groups of three.<br><br>**Further learning required**<br>• May be unable to represent and solve the problem by making equal groups, or arrives at the correct answer without fully understanding the problem, e.g. may share 12 cookies between 3 boxes and answer '4' because there are four in each box. |
| | 5i | 4 sweets each | **On track**<br>• Represents the problem by drawing or with concrete materials to show 16 items shared equally between four children, giving them four each.<br><br>**Further learning required**<br>• May not understand what 'share equally' means. |
| **Money**<br>MNU 1-09a<br>MNU 1-09b | 6a | Dog:<br>4 × 1p<br>2p 2p<br><br>Giraffe:<br>10p, 1p, 2p<br>13 × 1p<br>6 × 2p, 1p<br>5p, 5p, 2p, 1p<br><br>Elephant:<br>5p, 5p, 5p, 1p<br>10p, 5p, 1p<br>16 × 1p<br>8 × 2p<br><br>Penguin:<br>3 × 5p, 2p, 2p<br>10p, 5p, 2p, 2p<br>19 × 1p<br>9 × 2p, 1p<br><br>Other answers are possible. For example, the dog can also be paid for with 1 x 2p coin and 2 x 1p coins. | **On track**<br>• Correctly uses coins to make specified amounts of money.<br>• Understands that the same amount can be made in several ways using different coins.<br><br>**Further learning required**<br>• May not know/recognise all of the coins to 20p.<br>• May be unable to add a string of numbers or lacks a strategy to keep track of the calculation. |

| Topic – benchmarks / Es & Os | Question | Answer | Notes |
|---|---|---|---|
| | 6b | 14p<br>20p | **On track**<br>• Correctly adds amounts of money to work out how much to pay.<br><br>**Further learning required**<br>• May be unable to add a string of numbers or lacks a strategy to keep track of the calculation. |
| | 6c | Draws coins to represent the following amounts:<br>7p<br>5p<br>18p<br>10p | **On track**<br>The pupil may use a range of strategies to solve this problem including:<br>• Counting on from the cost of each item up to 20p, using coins, a number line or known facts.<br>• Subtracting the cost of each item from 20p, using coins, counting back on a number line or known facts.<br><br>**Further learning required**<br>• May be unable to select the correct operation to solve the problem; for example, adds each amount to 20p instead of counting on or subtracting.<br>• May select the correct operation but make calculation errors; for example, by using a number line incorrectly. |
| | 6d | 20p<br>Draws a 20p coin<br><br>14p<br>Draws a 10p coin and two 2p coins<br><br>7p<br>Draws a 5p and a 2p<br><br>16p<br>Draws a 10p coin, a 5p coin and a 1p coin | **On track**<br>• Can choose an appropriate strategy to add the amounts accurately.<br>• Recognises all coins to 20p and can use them to correctly make each total using the least number of coins.<br><br>**Further learning required**<br>• May not know all the coins to 20p.<br>• May make calculation errors; for example, by using a number line incorrectly.<br>• May make the correct amounts but not with the least number of coins; for example, makes 7p with seven 1p coins. |
| **Time**<br>MNU 1-10a<br>MNU 1-10b<br>MNU 1-10c | 7a<br>7b<br><br>7c | Answers will depend on the day on which the child is sitting the assessment<br>Wednesday at 1o'clock (or 1.00–2.00)<br>Thursday<br>1 hour | **On track**<br>• Correctly identifies the days of the week.<br>• Understands what today and yesterday mean.<br><br>**Further learning required**<br>• May not know the days of the week.<br>• May not understand what today and yesterday mean. |

| Topic – benchmarks / Es & Os | Question | Answer | Notes |
|---|---|---|---|
| | | | **On track** <br>• Can accurately read a simple timetable. <br><br>**Further learning required** <br>• May be unable to use a simple timetable. |
| | 7d | Lunch – minutes <br>Flower – days <br>Jump – seconds <br>Marathon – hours | **On track** <br>• Chooses the most appropriate unit of time for each example. <br><br>**Further learning required** <br>• May not understand standard units of time and the different duration each represents. |
| | 7e <br>7f | 2 o'clock <br>Half past 5 | **On track** <br>• Correctly uses o'clock and half past to record times shown on analogue clocks. <br><br>**Further learning required** <br>• May not understand standard units of time and the vocabulary o'clock and half past. |
| | 7g <br>7h | Half past 11 <br>Two o'clock | **On track** <br>• Correctly uses o'clock and half past to record times shown on digital clocks. <br><br>**Further learning required** <br>• May not be able to use a digital clock. <br>• May not understand standard units of time and the vocabulary o'clock and half past. |
| **Measurement** <br>MNU 1-11a <br>MNU 1-11b | 8a | Tallest – tower <br>Shortest – pencil <br>From shortest to tallest: <br>Pencil <br>Finlay <br>Car <br>Mum <br>Giraffe <br>Tower | **On track** <br>• Can identify the tallest and shortest items. <br>• Orders the items correctly from shortest to tallest. <br><br>**Further learning required** <br>• May not understand the vocabulary in the problem. <br>• May present their solution in the reverse order. |
| | 8b <br><br><br><br>8c | Pencil, because the scale is higher on the side with the pencil <br><br>Car, because the scale is lower on the side with the car | **On track** <br>• Understands what heavier and lighter mean. <br>• Can explain that the scale is lower down (towards the surface) when the item is heavier, and the scale is higher (further away from the surface) when the item is lighter. <br><br>**Further learning required** <br>• May not understand the vocabulary in the problem. <br>• May not understand how a balance works. <br>• May present their solution in the reverse order. |

| Topic – benchmarks / Es & Os | Question | Answer | Notes |
|---|---|---|---|
| | 8d | Finlay – smallest pool<br>Nuria – biggest pool | **On track**<br>• Understands that biggest means bigger than all of the others and smallest means smaller than all of the others.<br><br>**Further learning required**<br>• May not understand the vocabulary in the problem. |
| | 8e<br>8f | Jug<br>Smaller bin ticked | **On track**<br>• Understands that 'holds more' means has a larger capacity and 'holds less' means has a smaller capacity.<br><br>**Further learning required**<br>• May not understand the vocabulary in the problem.<br>• May be unable to visualise capacity in a 2D representation. |
| | 8g<br>8h | About 5 blocks<br>About 9 blocks<br><br>Accept reasonable alternative answers. You may wish to question the child to elicit how they have reasoned about the problems. | **On track**<br>• Understands the requirements of the questions and can give a reasonable estimate of height and length by comparing each picture with the non-standard unit given, i.e. the 'block'.<br><br>**Further learning required**<br>• May not understand the requirements of the task, e.g. may mistakenly think they have to compare real objects.<br>• May be unable to choose an appropriate strategy for comparing the objects. |
| | 8i | 5 beanbags | **On track**<br>• Understands the requirements of the question and counts the beanbags accurately.<br><br>**Further learning required**<br>• May not understand the requirements of the question.<br>• May count the beanbags incorrectly. |
| | 8j<br>8k | About 9 slices of toast<br>About 4 bin bags<br><br>Accept reasonable alternative answers. You may wish to question the child to elicit how they have reasoned about the problems. | **On track**<br>• Can make reasonable estimates using the images presented.<br><br>**Further learning required**<br>• May not have a strategy for comparing non-standard units of measurement. |

| Topic – benchmarks / Es & Os | Question | Answer | Notes |
|---|---|---|---|
| **Mathematics, its impact on the world, past, present, future** MTH 1-12a | 9 | Multiple answers are possible, for example: Supermarket Bakery Butcher Smartphone Games | **On track** • Knows that numbers are all around us and can give examples of how they are used in real life. **Further learning required** • May lack the experience of seeing numbers used in real-life contexts. |
| **Patterns and relationships** MTH 1-13a MTH 1-13b | 10a 10b 10c 10d | Pupil draws a square followed by a circle Pupil draws two triangles 14, 16 19, 23 | **On track** • Can identify and continue the shape patterns. • Can identify and continue the number patterns. **Further learning required** • May not understand that patterns can be made with shapes or numbers. • May be able to continue an a-b-a-b pattern (question a) but be confused by a-b-b-a pattern (question b). • May not have a strategy for solving number patterns, e.g. looking for the 'gap' between numbers. |
| **Expressions and equations** MTH 1-15a MTH 1-15b | 11a 11b 11c 11d | 5 8 – + | **On track** • Understands that the equal sign is used to show balance. • Uses known facts or the inverse relationship to solve the problems. **Further learning required** • May choose the wrong operation/strategy, e.g. answers 25 for question a) and 20 in question b). • May not realise that an equation can be 'solved' by inserting a symbol. |
| **2D shapes and 3D objects** MTH 1-16a MTH 1-16b | 12a 12b 12c | Draws a square and a rectangle Draws a triangle Draws a circle | **On track** • Knows the properties of common 2D shapes (square, rectangle, triangle, circle) and can make a recognisable drawing of each. **Further learning required** • May be unable to distinguish between a square and a rectangle. • May be unfamiliar with the properties of common 2D shapes and be unable to make a recognisable drawing of each. • May draw rectilinear shapes with curved lines or curved corners. |

| Topic – benchmarks / Es & Os | Question | Answer | | Notes |
|---|---|---|---|---|
| | 12d | 4<br>2<br>4 | | **On track**<br>• Understands what a corner is.<br>• May name the shapes and can sort them according to the criteria given, i.e. number of corners.<br><br>**Further learning required**<br>• May not know the vocabulary of 'corners'. |
| | 12e | i<br><br>ii | Cube, cuboid<br><br>Cylinder, sphere, cone | **On track**<br>• Correctly identifies 3D objects by number of faces.<br>• Understands what curved means and can identify the objects with at least one curved face.<br><br>**Further learning required**<br>• May not understand the language of the question, i.e. faces; curved.<br>• May have difficulty counting the faces of a 3D object from a 2D image. |
| **Angles, symmetry and transformation**<br><br>MTH 1-17a<br>MTH 1-18a<br>MTH 1-19a | 13a | Doll<br>Car<br>Tablet<br>Nothing<br>Robot<br>Football | | **On track**<br>• Understands the words left, right, above, below and can use this knowledge to identify the position of the objects in relation to each other.<br><br>**Further learning required**<br>• May confuse left and right.<br>• May be unfamiliar with positional language. |
| | 13b | Library<br>Hospital<br>Bank | | **On track**<br>• Understands the commands 'forwards, backwards, left, right' and can use this to correctly locate places on the plan.<br><br>**Further learning required**<br>• May be unfamiliar with the language used in the question<br>• May confuse left and right. |
| | 13c<br><br><br>13d | <br><br> | | **On track**<br>• Completes each pattern to make it symmetrical.<br><br>**Further learning required**<br>• May not understand the word 'symmetry'.<br>• May copy the patterns but not reflect them. |

| Topic – benchmarks / Es & Os | Question | Answer | Notes |
|---|---|---|---|
| **Data handling and analysis**<br>MNU 1-20a<br>MNU 1-20b | 14a | Tally column completed showing correct tallies for: 8, 6, 2, 5, 2<br><br>Total column completed with the above numbers. | **On track**<br>• Works systematically and correctly uses tallies to collect and record data.<br>**Further learning required**<br>• May not be able use tallies or know to complete totals.<br>• May record tallies without a 'gate' at five. |
| | 14b | Crisps | **On track**<br>• Understands the phrase 'most popular'.<br>**Further learning required**<br>• May be unfamiliar with the language of the question. |
| | 14c | Strawberry<br>8<br>2 | **On track**<br>• Counts the pictures accurately to find out which one has more.<br>• Knows that to find how many fewer they need to subtract/find the difference.<br>**Further learning required**<br>• May be unfamiliar with the language of the question and/or be unable to choose a strategy to find 'how many fewer'. |
| | 14d | Finlay has more brothers than sisters. | **On track**<br>• Correctly interprets data on a graph.<br>**Further learning required**<br>• May misinterpret the problem or the images on the graph. |
| **Ideas of chance and uncertainty**<br>MNU 1-22a | 15 | Possible<br>Impossible<br>Answer to the final question will depend on whether or not the child has a dog | **On track**<br>• Understands the meaning of the words, certain, possible and impossible.<br>• Can use impossible, certain and impossible to predict the likelihood of an event and justify their answer.<br>**Further learning required**<br>• May not understand the mathematical interpretation of the chance words, e.g. refusing to use 'impossible' in the belief that 'nothing is impossible'. |

# Yearly progress check 1B

## Questions

1. a

```
40  41  42  43  44  45  46  47  48  49  50
```

Is 44 closer to 40 or 50?

44 rounded to the nearest ten is ▢.

b

```
80  81  82  83  84  85  86  87  88  89  90
```

Is 88 closer to 80 or 90?

88 rounded to the nearest ten is ▢.

c Finlay says that 65 rounded to the nearest ten is 70. Nuria thinks it is 100.

Who is right? _____

Explain your thinking.

d Estimate the answer to this problem using rounding to the nearest ten.

38 + 13

▢ + ▢ = ▢

2. **67**

a Write this number in words. _____

b Swap the order of the digits in the number 67. Write the new number in words.

_____  _____

# Yearly progress check 1B

c  Fill in the missing numbers

| 300 |  | 500 |  | 700 |  |  |
|-----|--|-----|--|-----|--|--|

d  Write any three-digit number in numerals.

_____

Now write it in words.

_____

e  Find the missing numbers and write them in the boxes.

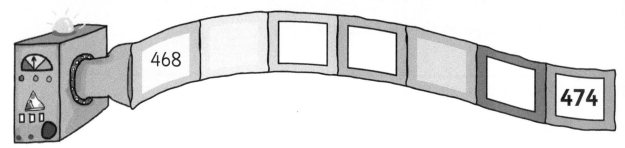

f  Find the missing numbers and write them in the boxes.

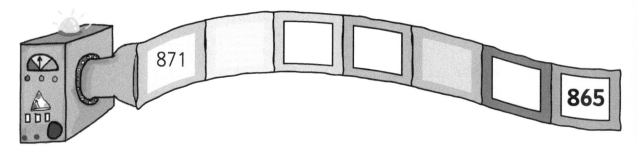

g  Write the number that comes after:

699 _____          480 _____

h  Write the number that comes before:

_____ 601          _____ 299

i  Write all the numbers that come in-between:

236 and 246

_____

907 and 917

_____

j   Count forward in tens to find the missing numbers and write them in the boxes.

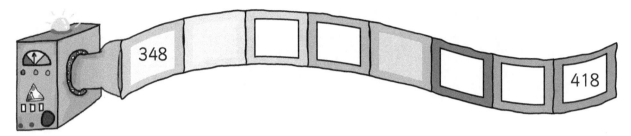

348 ☐ ☐ ☐ ☐ ☐ ☐ 418

k   Count forward in hundreds to find the missing numbers and write them in the boxes.

66 ☐ ☐ ☐ ☐ ☐ ☐ 766

l   Read this number **678**. Now write the number that is:

1 less

_____

10 less

_____

100 less

_____

m   How much money does Nuria have?

_____

n   How much money does Findlay have?

_____

o   Count in hundreds and tens.

How many dots altogether?

_____

p   Count in hundreds, tens and ones.

How many dots altogether?

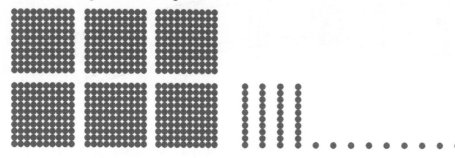

_____

q   Match each ball to a goal:

68

15

51

45

60

40 + 5

50 + 1

60 + 0

60 + 8

10 + 5

r   Finlay has 64 stickers and Nuria has 46.

Who has more? _____

Explain your thinking? _____

_____

_____

s   Choose the correct phrase and write it in the box:

is greater than;        is equal to;        OR        is less than

501 | [                    ] | 675 | [                    ] | 675

t   Write these numbers from smallest to largest.

357, 360, 341, 305, 399 _____

u   Write these numbers from largest to smallest.

580, 340, 490, 930, 760 _____

v

Ted   Kim   Steve   Kate   Mike   Paul   Sue   Jack   Jen   Tom   Jim   Dave   Liz   Bill

In what positions are these drivers? Write your answers in numerals and words.

Kim _____

Jim _____

Mike _____

Ted _____

3.  Find each total. Explain how you worked them out.

a  8 + 6 + 2 = _____

b  3 + 4 + 7 = _____

c  6 + 5 = _____

d  8 + 9 = _____

e

| 19 | |
|---|---|
| 12 | 7 |

Write a fact family for the bar model.

_____

_____

_____

_____

f  56 + ☐ = 60

g  67 − ☐ = 60

h  Find the missing numbers for each function machine.

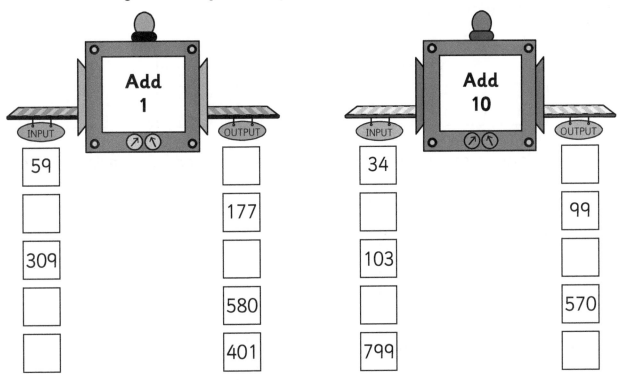

i  Find the missing numbers for each function machine.

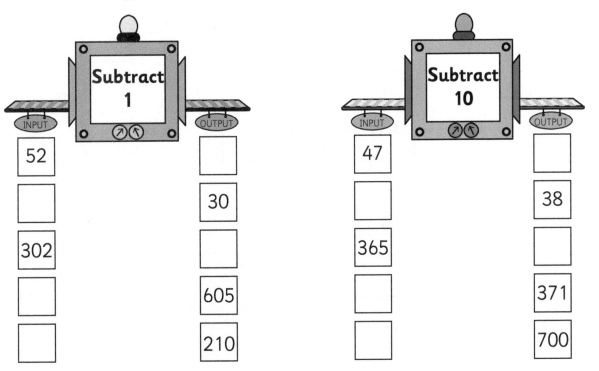

# Yearly progress check 1B

j Complete the additions, then write the next three number sentences to continue the pattern.

3 + 5 = ☐          13 + 5 = ☐          23 + 5 = ☐

☐ + ☐ = ☐          ☐ + ☐ = ☐          ☐ + ☐ = ☐

k Complete these number sentences. Write the addition facts you used to help you.

4 + 35 = ☐

7 + 72 = ☐

l Fill in the missing numbers.

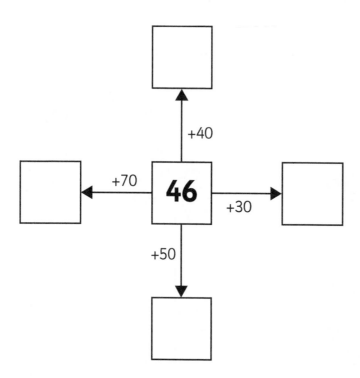

m  Fill in the missing numbers.

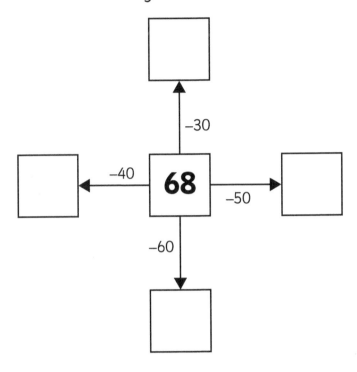

n  Solve these additions and subtractions. Show your thinking on empty number lines.

67 + 8 = [ ]

_____

48 + 5 = [ ]

_____

95 − 7 = [ ]

_____

76 − 7 = [ ]

_____

o   Solve these problems by partitioning the numbers into tens and ones.

34 + 13 = _____

17 + 57 = _____

78 – 31 = _____

88 – 33 = _____

p   Draw empty number lines to solve:

79 + [          ] = 97

40 = 35 + [          ]

q   Write one addition number sentence and one subtraction number sentence for this number line.

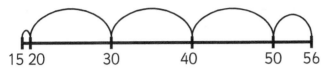

15 20      30      40      50   56

r   Write a number in the empty circles to make each line have the same total.

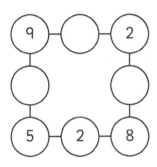

s   Complete a Think Board for each question.

   i   The pet shop has 19 hamsters and only 11 cages. If hamsters live alone, how many hamsters won't have a cage?

| Bar model | Objects / picture / diagram |
|---|---|
|  |  |
| Number sentence and answer | Empty number line |
|  |  |

# Yearly progress check 1B

ii Nuria has 20 dolls and some cars. She has 4 fewer cars than dolls. How many cars does Nuria have?

| Bar model | Objects / picture / diagram |
|---|---|
| **Number sentence and answer** | **Empty number line** |

iii Finlay has been playing a computer game. He scored 65 points today. Yesterday he scored 38. How many more points did Finlay score yesterday?

| Bar model | Objects / picture / diagram |
|---|---|
| **Number sentence and answer** | **Empty number line** |

iv  Mrs Jones sold 56 sweets at the tuck shop. There are only 17 sweets left. How many sweets did she start with?

| Bar model | Objects / picture / diagram |
|---|---|
| Number sentence and answer | Empty number line |

t  Write two addition number sentences and two subtraction number sentences to fit this bar model. Write a story problem about the bar model.

| 19 | |
|---|---|
| 11 | 8 |

_____

_____

_____

_____

4. a  Isla has made cakes. She has 8 boxes and she put 3 cakes in each box.

   How many cakes did she make altogether? Use an empty number line to show your thinking.

   b  Write these number problems as additions, then skip count to find the answers:

   There are four football teams. Each team has five players. How many players in total?

   There are seven pencil cases. Each pencil case has three pencils. How many pencils are there in total?

# Yearly progress check 1B

c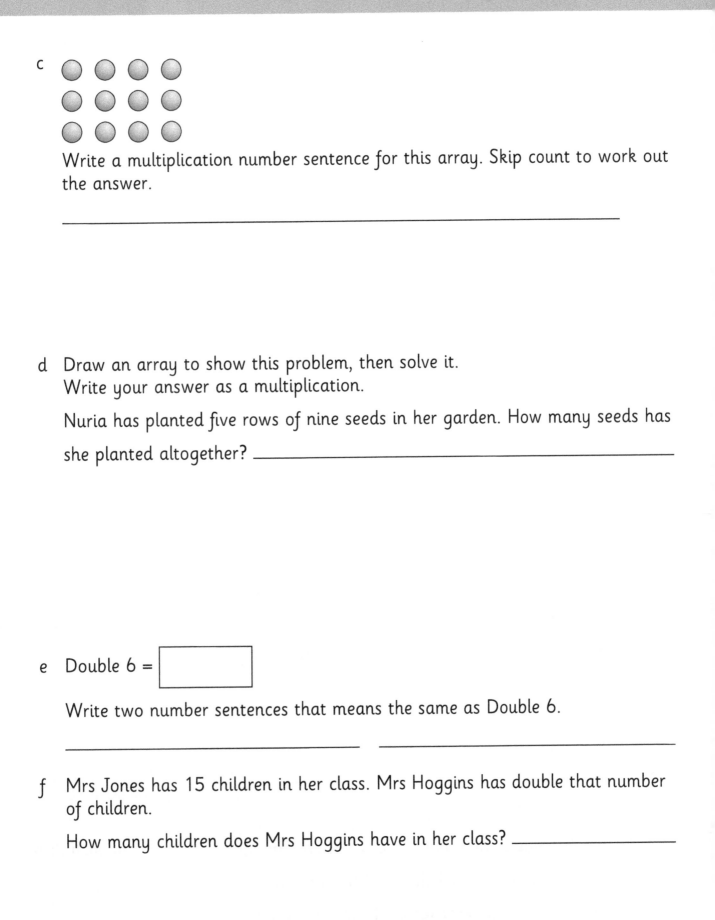

Write a multiplication number sentence for this array. Skip count to work out the answer.

_____

d Draw an array to show this problem, then solve it.
Write your answer as a multiplication.

Nuria has planted five rows of nine seeds in her garden. How many seeds has

she planted altogether? _____

e Double 6 = ▢

Write two number sentences that means the same as Double 6.

_____   _____

f Mrs Jones has 15 children in her class. Mrs Hoggins has double that number
of children.

How many children does Mrs Hoggins have in her class? _____

g   Share 35 dog chews equally between five dogs.

How many dog chews will each dog get? _____

You may use counters or draw a picture to help you.

h   Amman shares 24 pieces of paper equally between 6 tables.

How many pieces of paper will each table get? _____

You may use counters or draw a picture to help you.

i   Finlay has made 36 brownies and wants to put 6 in each box.

How many boxes will he need? _____

You may use counters or draw a picture to help you.

j   Isla is feeding the rabbits in a pet shop. She puts five carrots in each dish.

How many dishes will she need for 45 carrots? _____

Write a number sentence to show how you worked this out.

_____

k   Maggie went to the bank with her mum. She had £80 in £10 notes.

How many £10 notes did she have? _____

Write a number sentence to show how you worked it out.

_____

l   How many tens on the abacus? Write a division number sentence to match the picture.

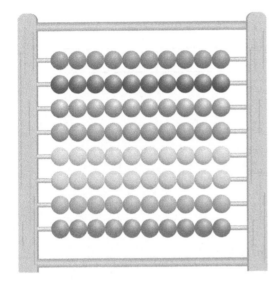

5. a   Circle the shapes that have been split into equal parts.

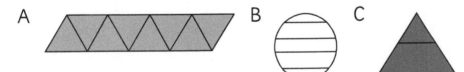

A      B      C      D

b   What fraction of the rectangle has been shaded?

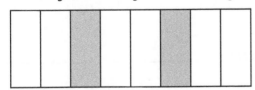

c   Some of each pizza has been eaten.

What fraction is left on each plate?

d  Show these fractions on the bars and write them in numbers.

| | | | | | |
|---|---|---|---|---|---|

Three sixths =

| | | | | | |
|---|---|---|---|---|---|

One third =

| | | | | |
|---|---|---|---|---|

Two fifths =

| | | | | | | |
|---|---|---|---|---|---|---|

Three quarters =

e  Which is larger, three fifths or three quarters? _____

Draw a diagram to show your answer.

f  Which is larger, three tenths or four sixths? _____

Draw a diagram to show your answer.

g   Draw a number line that counts on from four to ten in jumps of one half.

h   Finlay had £45. He spent one fifth of his money on a book about cars.

How much did Finlay's book cost? _____

i   Amman had 28 pencils. He lost one quarter of them.

How many pencils did Amman lose? _____

j   Nuria planted 36 seeds. One third of the seeds did not grow.

How many seeds did not grow? _____

6.  Write down the total amounts in pounds and pence.

a  _____

b

_____

c

_____

Write down the total amounts in pence.

d   £5 note, a 50p and a 20p

_____

e   £1 coin, £2 coin, 10p, 20p, 1p, 2p

_____

f   £8.79

_____

Now try these money problems. Show how you worked each answer out.

g   Isla orders a burger that costs 50p and a milkshake that costs 23p.

How much money does Isla need? _____

h Finlay has 65p. He is in the shop and would like crisps costing 45p and juice costing 36p.

How much more money does Finlay need to be able to buy both things?

_____

i Lucy has £1. She spends 65p on stickers.

How much change does she get? _____

j Olivia has £1. She spends 48p on crisps.

How much change does she get? _____

k Maggie has a £2 coin. She spends £1.25 on a pencil.

How much change does she get? _____

# Yearly progress check 1B

7.  a  Finlay spent 15 minutes on his maths. He answered four questions altogether, but it took him 8 minutes to answer the first three questions.

How long did it take Finlay to do the last question? _____

b  It took Nuria 28 hours to fly to Australia. Amman was on a different flight and it took him 1 day to get to Australia.

Who got to Australia quicker? _____

Explain how you know?

_____

c  The builders took 98 months to build the supermarket. They then took 74 months to build the car park.

How much longer did it take to build the supermarket? _____

d  Amman's dad was working in Spain. He was away from Thursday to Wednesday.

How many days was he working? _____

e

JULY

| Sun | Mon | Tue | Wed | Thu | Fri | Sat |
|-----|-----|-----|-----|-----|-----|-----|
|     |     | 1   | 2   | 3   | 4   | 5   | 6 |
| 7   | 8   | 9   | 10  | 11  | 12  | 13  |
| 14  | 15  | 16  | 17  | 18  | 19  | 20  |
| 21  | 22  | 23  | 24  | 25  | 26  | 27  |
| 28  | 29  | 30  | 31  |     |     |     |

Mike and Lola went on holiday. They arrived at the hotel on 17th July and stayed for 12 nights.

What date did they leave the hotel? _____

f  How many months are there in a year? _____

g  How many months are there in four years? _____

8.  a  Use a ruler. Draw a line that is exactly:

4 cm long

18 cm long

b  Estimate and then measure how many sticky notes will fit onto this piece of paper without overlapping or leaving any gaps. _____

c  Estimate and then measure how many pieces of paper will fit onto the desk you are working at without overlapping or leaving any gaps. _____

d  Is it better to estimate the capacity of the mug in litres or millilitres?

_____

Is it better to estimate the capacity of the pool in litres or millilitres?

_____

Explain your answers.

_____

_____

_____

_____

# Yearly progress check 1B

e   Write down how tall each object is in metres and centimetres.

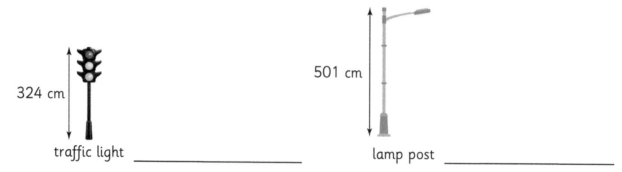

324 cm

traffic light _____

501 cm

lamp post _____

f   Find the total mass of these weights in kilograms and grams.

_____ kg _____ g

g   Find the total mass of these weights in kilograms and grams.

_____ kg _____ g

Use the ruler to measure the items below.

h

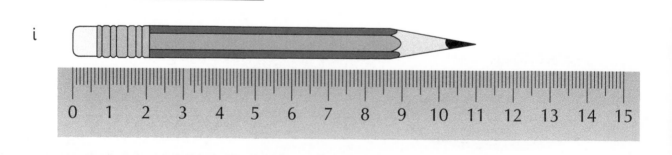

_____

i

_____

j   What is the mass of the sugar?

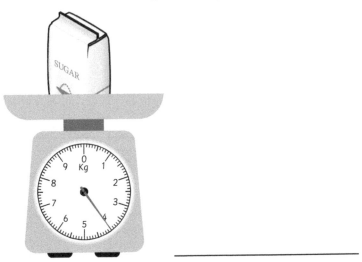

_____

k   What mass is the scale showing?

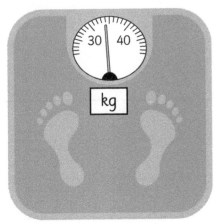

_____

l   How much water is in this measuring jug?

_____

# Yearly progress check 1B

9.  a  Write two ways nurses might use maths in their job.

_____

_____

_____

_____

b  Write two ways bakers might use maths in their job.

_____

_____

_____

_____

10.  a  Continue the shape pattern.

b  Complete these number patterns

3, 6, 9, 12, [       ], [       ], [       ].

40, 50, [       ], 70, [       ], 90

# Yearly progress check 1B

11. Solve the following equations:

a  15 + ☐ = 30

b  50 − ☐ = 25

c  ☐ + 16 = 25

12. a

square    rectangle    circle    hexagon    pentagon    right-angled    semi-circle
                                                         triangle

Which shapes have 4 corners? _____

Which shapes have more than 4 sides? _____

Which shape has only one straight edge? _____

Which shapes have fewer than 4 sides? _____

_____

Which shapes have curved sides? _____

Which shape has only one right angle? _____

b  Finlay is describing 3D objects. Name the objects he is describing.

i   It has 6 vertices and 9 edges. _____

ii  It has 12 edges and 8 vertices. _____

c  What shape is the box?

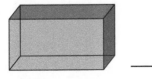  _____

d  Name three 2D shapes that tile.

_____

13. a  Tick the shape that has made a quarter turn.

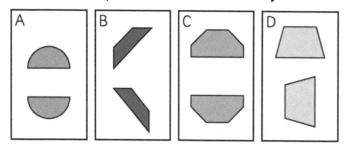

b  Tick the shape that has made a quarter turn.

c

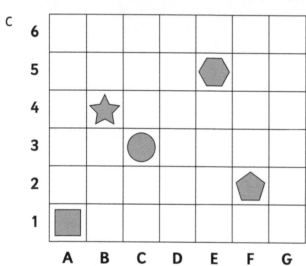

Write down the grid reference for each shape.

Square _____

Circle _____

Hexagon _____

Pentagon _____

Star _____

d   Tick the symmetrical shapes.

Cross out the shapes that are not symmetrical.

14.  a   Look at the tally chart. It shows a tally of favourite toys.

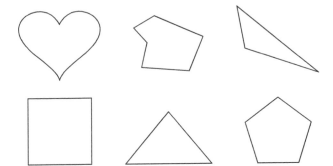

| Toy | Tally |
|---|---|
| (doll) | ⊪⊦ \| |
| (ball) | ⊪⊦ ⊪⊦ |
| (bear) | ⊪⊦ \|\|\|\| |
| (car) | ⊪⊦ |

i   Which toy is the most popular?

_____

ii   How many children's favourite toy was a doll?

_____

iii   How many children's favourite toy was a toy car?

_____

iv   Which toy was the least favourite?

_____

# Yearly progress check 1B

b   Read the information and complete the picture graph. Use this symbol ☺ for your picture graph

5 children have hazel eyes

8 children have blue eyes

3 children have green eyes

9 children have brown eyes

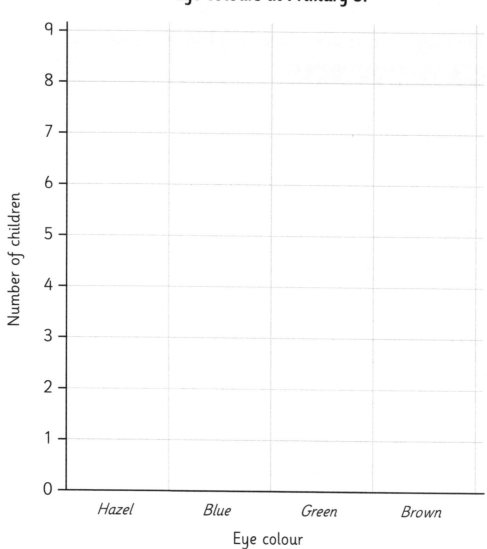

**Eye colours in Primary 3.**

c  **How P3 travel to school**

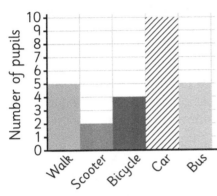

Look at the graph and answer these questions:

i   Which method of transport is most popular?

_____

ii  Which methods of transport got the same number of votes?

_____

iii Which method of transport is least popular?

_____

iv  Which method of transport got 2 more votes than scooter?

_____

v   How many more pupils travel to school by car than by bus?

_____

15. Answer using **Impossible, Unlikely, Likely, Certain.** Write your answer next to each statement:

I will be older tomorrow than I am today. _____

I will be in school on Saturday. _____

I will meet an alien. _____

It will rain cats and dogs. _____

| Topic – benchmarks / Es & Os | Question | Answer | Notes |
|---|---|---|---|
| **Estimation and rounding**<br><br>MNU 1-01a | 1a<br>1b<br>1c | 40<br>90<br>70 Finlay is correct because he has rounded to the nearest 10. Nuria has rounded to the nearest 100. | **On track**<br>• Responds correctly using the number line provided or by visualising the number line in their mind's eye.<br>• Uses knowledge of reading and writing decade numbers to facilitate rounding.<br><br>**Further learning required**<br>• May not understand the meaning of the word closer nor know the number sequence from 0 to 9 on either side of a two-digit number.<br>• May not understand what each digit in the number represents.<br>• May round incorrectly for example, thinks Nuria is correct in question c because they see '10' in the number 100. |
| | 1d | 40 + 10 = 50 | **On track**<br>• Correctly rounds to find the approximate answer.<br><br>**Further learning required**<br>• May be unable to read and write decade numbers to 100.<br>• May not understand what each digit in the numbers represent.<br>• May round incorrectly, e.g. 30 + 10.<br>• May be able to round 38 and 13 to the nearest 10, but be unable/forget to add the decade numbers to solve the problem. |
| **Number – order and place value**<br><br>MNU 1-02a | 2a | Sixty-seven | **On track**<br>• Writes the number in words, making a reasonable attempt to spell the words correctly.<br><br>**Further learning required**<br>• May not know the number words to use. |
| | 2b | 76<br>Seventy-six | **On track**<br>• Correctly reverses the digits.<br>• Writes the new number in words correctly, making a reasonable attempt to spell the words correctly.<br><br>**Further learning required**<br>• May reverse the digits, but be unable to write the number in words. |

| Topic – benchmarks / Es & Os | Question | Answer | Notes |
|---|---|---|---|
| | 2c | 400, 600, 800, 900 | **On track**<br>• Completes the number sequence correctly, counting in 100s.<br>• Correctly reads and writes the numbers.<br><br>**Further learning required**<br>• May be unsure of the patterns in the way we say numbers.<br>• May count in 1s or 10s instead of 100s. |
| | 2d | Pupil's own choice of 3-digit number, written correctly in numerals and words. | **On track**<br>• Writes their chosen number in words, making a reasonable attempt to spell the words correctly.<br>• Writes the numerals correctly with the digits in the correct order, i.e. they 'match' the words.<br><br>**Further learning required**<br>• May not know the number words to use.<br>• May choose the wrong number words, e.g. creates the number 145, but writes one hundred and fifty-four in words. |
| | 2e | 469, 470, 471, 472, 473 | **On track**<br>• Continues the number sequence correctly, counting forwards in 1s.<br><br>**Further learning required**<br>• May count incorrectly or have difficulty bridging a 10.<br>• May count in 100s or 10s instead of 1s. |
| | 2f | 870, 869, 868, 867, 866 | **On track**<br>• Continues the number sequence correctly, counting backwards in 1s.<br><br>**Further learning required**<br>• May count incorrectly or have difficulty bridging a 10.<br>• May count in 100s or 10s instead of 1s. |
| | 2g | 700<br>481 | **On track**<br>• Correctly identifies the number after.<br>• Correctly identifies the number before.<br>• Correctly identifies the numbers that come in between. |
| | 2h | 600<br>298 | |
| | 2i | 237, 238, 239, 240, 241, 242, 243, 244, 245<br>908, 909, 910, 911, 912, 913, 914, 915, 916 | **Further learning required**<br>• May not understand the language of before, after and in between.<br>• May be unable to bridge a 10 or 100.<br>• May be able to count forwards but find counting backwards more challenging. |

| Topic – benchmarks / Es & Os | Question | Answer | Notes |
|---|---|---|---|
| | 2j | 358, 368, 378, 388, 398, 408 | **On track**<br>• Counts forwards correctly in 10s.<br><br>**Further learning required**<br>• May count in 1s or 100s instead of 10s.<br>• May have difficulty bridging a 10/100, e.g. moving from 398 to 408. |
| | 2k | 166, 266, 366, 466, 566, 666 | **On track**<br>• Counts forwards correctly in 100s.<br><br>**Further learning required**<br>• May count in 1s or 10s instead of 100s. |
| | 2l | Reads the number 678 correctly<br>677<br>668<br>578 | **On track**<br>• Reads the number 678 correctly.<br>• Appreciates the value of the digits and can therefore work out 1, 10 or 100 less without having to count.<br><br>**Further learning required**<br>• May say the number incorrectly, e.g. "six, seven, eight" or "sixty-seven eight".<br>• May need to first count forwards to find one less, e.g. counts '664, 665, 666, 667, 668'.<br>• May rely on counting backwards in 1s to find 10 less.<br>• May be unable to identify the number that is 100 less as a result of the aforementioned difficulties. |
| | 2m<br>2n | 28p<br>100p or £1 | **On track**<br>• Counts forwards and backwards in twos and fives correctly.<br><br>**Further learning required**<br>• May give answers of 14p and 20p because there are 14 and 20 coins respectively.<br>• May have difficulty counting in 2s and 5s to find 'how many' where the individual pennies with the equivalent value of each 2p or 5p cannot be seen.<br>• May be unable to count in 2s beyond 20 and/or in 5s beyond 50. |
| | 2o<br>2p | 570<br>649 | **On track**<br>• Correctly counts in hundreds, tens and ones to arrive at the correct totals.<br><br>**Further learning required**<br>• May be unsure of the required number sequences and so attempted to count the dots in ones. |

| Topic – benchmarks / Es & Os | Question | Answer | Notes |
|---|---|---|---|
| | | | • May be able to count in hundreds, tens or ones independently (e.g. counts 600, 40 and 9 in question p) but is unable to combine these 'counts' to find the total.<br>• May arrive at the correct totals but be unable to record these in numerals or records the total incorrectly, e.g. writes five hundred and seventy as 50070 or 507. |
| | 2q | 68 = 60 + 8<br>15 = 10 + 5<br>51 = 50 + 1<br>45 = 40 + 5<br>60 = 60 + 0 | **On track**<br>• Correctly partitions two-digit numbers into tens and ones.<br>**Further learning required**<br>• May be unable to see 68, for example, as anything other than 6 tens and 8 ones and so has difficulty completing the task.<br>• May not appreciate the values of the digits in a 2-digit number and so matches the balls and goals incorrectly, e.g. thinks 51 = 10 + 5. |
| | 2r | Finlay, because he has 64 (60 + 4) which is more than 46 (40 + 6) | **On track**<br>• Correctly compares numbers by using place value and can explain, or show using concrete materials or a number line, why Finlay has more.<br>**Further learning required**<br>• May lack confidence in reading 2-digit numbers and/or mistakenly believe that 46 is more than 64 because there are more ones.<br>• May be unable to explain or show why Finlay has more. |
| | 2s | Is less than<br>Is equal to | **On track**<br>• Correctly uses less than, equal to and greater than to compare numbers.<br>**Further learning required**<br>• May not understand what less than and greater than mean.<br>• May read numbers in the wrong order, e.g. using the ones value to compare size. |
| | 2t<br>2u | 305, 341, 357, 360, 399<br>930, 760, 580, 490, 340 | **On track**<br>• Orders numbers from smallest to largest and largest to smallest correctly. |

| Topic – benchmarks / Es & Os | Question | Answer | Notes |
|---|---|---|---|
| | | | **Further learning required**<br>• May not understand the language of largest and smallest.<br>• May only be able to order numbers with the same number of hundreds, i.e. question t<br>• May believe the numbers in question u are of equivalent value since they all have zero in the ones place.<br>• May attempt to order individual digits. |
| | 2v | Kim – 2nd/second<br>Jim – 11th/eleventh<br>Mike – 5th/fifth<br>Ted – 1st/first | **On track**<br>• Uses ordinal numbers to correctly describe each driver's position.<br>**Further learning required**<br>• May not understand the language of ordinal numbers.<br>• May miscount or reverse the driver's positions. |
| **Number – addition and subtraction**<br>MNU 1-03a | 3a<br>3b | 16<br>14 | **On track**<br>• Knows the numbers can be added in any order and calculates mentally by looking for pairs that add to 10, e.g. 8 + 6 + 2 becomes 8 + 2 (10) + 6; 3 + 4 + 7 becomes 7 + 3 (10) + 4.<br>• May use doubles, e.g. 8 + 6 +2 might be found by thinking 6 +2 = 8; 8 + 8 = 16. Likewise, 3 + 4 = 7; 7 + 7 = 14.<br>• Can explain how they worked the answer out, orally or in writing.<br>**Further learning required**<br>• May not use these strategies and continues to 'count all' by ones.<br>• May use one of the above strategies incorrectly.<br>• May be unable to explain or show how they worked the answer out |
| | 3c<br>3d | 11<br>17 | **On track**<br>• Uses doubles and near doubles, e.g. finds 6 + 5 by doubling 5 and adding 1 or doubling 6 and subtracting 1.<br>• Uses known facts, e.g. knows that 9 is one less than 10 so reasons that the answer to 8 + 9 must be one less that 8 + 10.<br>• Can explain how they worked the answers out, orally or in writing. |

| Topic – benchmarks / Es & Os | Question | Answer | Notes |
|---|---|---|---|
| | | | **Further learning required**<br>• May rely on counting by ones, using concrete materials or a number line.<br>• May use one of the above strategies incorrectly.<br>• May be unable to explain or show how they worked the answers out. |
| | 3e | 19 – 7 = 12;<br>19 – 12 = 7;<br>7 + 12 = 19;<br>12 + 7 = 19 | **On track**<br>• Knows that a fact family is a set of four related facts (two additions and two subtractions) generated from the same whole and parts.<br>• Uses knowledge of 'part-part-whole' to create a fact family about the bar model.<br><br>**Further learning required**<br>• May be unfamiliar with the phrase 'fact family'.<br>• May write incorrect number sentences, e.g. 9 8 + = 17; 9 - 8 = 17.<br>• May be able to generate two additions from the bar model but find subtractions more difficult. |
| | 3f<br>3g | 4<br>7 | **On track**<br>• Uses counting on and back to a multiple of 10 to find the missing number.<br><br>**Further learning required**<br>• May not trust counting on and back as a strategy and attempt to count in 1s using concrete materials.<br>• May count on or back, incorrectly, for example by including the starting number (says 56, 57, 58, 59, 60 and so believes that 56 + 5 = 60 because they have said five numbers). |
| | 3h | 60<br>176<br>310<br>579<br>400<br><br>44<br>89<br>113<br>560<br>809 | **On track**<br>• Uses knowledge of place value, and the inverse relationship between addition and subtraction, to complete each function machine correctly.<br><br>**Further learning required**<br>• May have difficulty with examples which require them to bridge a 10, e.g. 799 + 10<br>• May rely on counting on in 1s to add 10.<br>• May not understand the inverse and so wrongly adds 1 or 10 to find missing input numbers. |

| Topic – benchmarks / Es & Os | Question | Answer | Notes |
|---|---|---|---|
| | 3i | 51<br>31<br>301<br>606<br>211<br><br>37<br>48<br>355<br>381<br>710 | **On track**<br>• Uses knowledge of place value, and the inverse relationship between addition and subtraction, to complete each function machine correctly.<br><br>**Further learning required**<br>• May have difficulty with examples which require them to bridge a 10, e.g. 700 – 10.<br>• May rely on counting back in 1s to subtract 10.<br>• May not understand the inverse and so wrongly subtracts 1 or 10 to find missing input numbers. |
| | 3j | 8<br>18<br>28<br>33 + 5 = 38<br>43 + 5 = 48<br>53 + 5 = 58 | **On track**<br>• Uses knowledge of number patterns and addition and subtraction facts to complete the first three additions.<br>• Uses place value knowledge (adding 10 off the decade) to extend the pattern.<br><br>**Further learning required**<br>• May not understand the value of the digits and/ or be unable to identify or extend the pattern.<br>• May make calculation errors. |
| | 3k | 4 + 35 = 39<br>5 + 4 = 9 or 4 + 5 = 9<br>7 + 72 = 79<br>7 + 2 = 9 or 2 + 7 = 9<br><br>Other explanations are possible, for example:<br>30 + 5 + 4 or 70 + 7 + 2 | **On track**<br>• Uses known facts and place value to complete each addition.<br><br>**Further learning required**<br>• May be unsure of number bonds.<br>• May have memorised number bonds to 10 but is unable to use these to add a single digit number to a 2-digit number.<br>• May count on in ones, using a number line or their fingers to keep track of the count.<br>• May rely on concrete materials to solve the problem. |
| | 3l<br><br><br><br>3m | 46 + 40 = 86<br>46 + 30 = 76<br>46 + 70 = 116<br>46 + 50 = 96<br>68 – 30 = 38<br>68 – 50 = 18<br>68 – 60 = 8<br>68 – 40 = 28 | **On track**<br>• Counts on and back in 10s to add and subtract decade numbers, perhaps using an empty number line to help them keep track of their count.<br>• Uses knowledge of place value, e.g. knows that if they are adding 30 to 46 they are adding 3 tens so the tens digit will change from 4 to 7.<br><br>**Further learning required**<br>• May be unable to use any of these strategies and attempts to solve the problems by counting in ones. |

| Topic – benchmarks / Es & Os | Question | Answer | Notes |
|---|---|---|---|
| | 3n | 75<br>53<br>88<br>69<br>Working out should be shown on empty number lines. | **On track**<br>• Uses partitioning through 10 on an empty number line to add or subtract a single digit number to or from a 2-digit number, e.g. calculates $67 + 8$ as $67 + 3 = 70 + 5 = 75$.<br><br>**Further learning required**<br>• May be unable to partition through 10 and continue to rely on counting by ones.<br>• May be unable to use an empty number line or does so incorrectly, counting numbers instead of jumps. |
| | 3o | **47;** $30 + 10 = 40$ and $4 + 3 = 7$<br>**74;** $10 + 50 = 60$ and $7 + 7 = 14$<br>**47;** $70 - 30 = 40$ and $8 - 1 = 7$<br>**55;** $80 - 30 = 50$ and $8 - 3 = 5$ | **On track**<br>• Uses a split strategy and partitioning into tens and ones to calculate the answer, e.g. knows that $34 = 30 + 4$ and $13 = 10 + 3$; $30 + 10 = 40$ and $4 + 3 = 7$ so the answer is 47.<br>• Correctly writes number sentences to show their thinking.<br><br>**Further learning required**<br>• May be unable to add tens and ones mentally and continue to rely on concrete materials.<br>• May be unsure of place value and so be uncertain about which digits to add together. |
| | 3p | 18<br>5<br><br>Empty number lines drawn to show how the pupils solved the problems. | **On track**<br>• Can use a jump strategy on an empty number line to solve missing addend problems involving two-digit numbers.<br>• Uses 5 and 10 as reference points, rather than counting ones.<br><br>**Further learning required**<br>• May not know what an empty number line is.<br>• May use the number line incorrectly, e.g. may include the starting number.<br>• May continue to count by ones on the number line.<br>• May misinterpret the problems and select the wrong operation/strategy, e.g. may add rather than continuing on/finding the difference. |
| | 3q | $56 - 41 = 15$<br>$15 + 41 = 56$ | **On track**<br>• Understands that a missing addend task can be solved by counting on or counting back.<br>• Can link counting on and counting back with addition and subtraction. |

| Topic – benchmarks / Es & Os | Question | Answer | Notes |
|---|---|---|---|
| | | | **Further learning required**<br>• May use the number line incorrectly, e.g. may include the starting number and so think the first jump, moving left to right has a value of 6. |
| | | | **Further learning required**<br>• May be able to work out and total the value of the 'jumps' but be unable to connect this to the appropriate number sentences. |
| | 3r | 1(left),4 (top), 5 (right) | **On track**<br>• Can reason about the problem and can identify a strategy for working out the value of each row and column, e.g. total bottom row first.<br>• Uses known facts and counting on and back to find missing values.<br>**Further learning required**<br>• May not appreciated that each row and column must total 15.<br>• May add the two known numbers in each row/column but not subtract the total from 15 to find the unknown number.<br>• May be unsure of the number bonds to 20 and/or be unable to identify a suitable strategy to help them. |
| | 3s | i  $11 + ? = 19$ or<br>$19 - ? = 11$ or<br>$19 - 11 = ?$<br>8 hamsters<br><br>ii  $20 - 4 = ?$ or<br>$4 + ? = 20$ or<br>$? + 4 = 20$<br>16 cars | **On track**<br>• Completes a Think Board for each problem by:<br>• Drawing a bar model which accurately represents the problem (part-part-whole).<br>• Drawing an empty number line and counting on or back to find the unknown. |

| Topic – benchmarks / Es & Os | Question | Answer | Notes |
|---|---|---|---|
| | | iii   $38 + ? = 65$ or $65 - 38 = ?$ <br> 27 points <br><br> iv   $? - 56 = 17$ or $56 + 17 = ?$ <br> 73 sweets | • Writing an appropriate number sentence that matches thinking/method (see answer column). <br> • May create a meaningful 'story' for the problem, e.g. by drawing a picture or acting the situation out with concrete materials. <br><br> **Further learning required** <br> • May be unable to make sense of each situation, particularly those involving larger numbers that are not easily visualised or represented with concrete materials. <br> • May be unable to represent each problem as a bar model. <br> • May make calculation or counting errors. <br> • May be unable to write a number sentence to match how they solved each problem. |
| | 3t | $19 - 11 = 8$ <br> $19 - 8 = 11$ <br> $8 + 11 = 19$ <br> $11 + 8 = 19$ | **On track** <br> • Can interpret the bar model in terms of 'part-part-whole' and write a family of four related facts from the numbers shown. <br> • Can create a story problem using the numbers in the bar model that shows understanding of 'part-part-whole', orally or in writing. <br><br> **Further learning required** <br> • May be unable to interpret the bar model. <br> • May be able to generate one or two addition number sentences but find subtractions more challenging. <br> • May write the number sentences incorrectly. <br> • May be unable to attach a suitable 'story' to the bar model. |
| **Number – multiplication and division** <br> MNU 1-03a | 4a | 24 cakes | **On track** <br> • Correctly identifies multiplication/repeated addition as the required operation. <br> • Skip counts in threes to solve the problem. <br> • Can show their thinking on an empty number line. <br><br> **Further learning required** <br> • May be unable to select the correct operation. <br> • May be unsure of the skip counting pattern in threes and resort to counting in ones, e.g. counts **3**, **6**, **9**,10, 11, **12**, 13, 14, **15** etc. |

| Topic – benchmarks / Es & Os | Question | Answer | Notes |
|---|---|---|---|
| | 4b | $5 + 5 + 5 + 5 = 20$<br>$6 + 6 + 6 + 6 + 6 + 6 + 6 = 42$ | **On track**<br>• Understands multiplication as repeated equal groups.<br>• Can recall the number word sequence in fives and sixes.<br><br>**Further learning required**<br>• May not be able to recall the number word sequences for both, e.g. can skip count in fives, but finds skip counting in sixes more challenging.<br>• May resort to counting in ones for all or part of the sequence. |
| | 4c | $3 \times 4 = 12$ | **On track**<br>• Can write a number sentence to match the array shown.<br>• Skip counts correctly in 3s or 4s to find the solution.<br><br>**Further learning required**<br>• May not understand what an array is.<br>• May interpret the array as '3 rows of 4' and/or $4 + 4 + 4$ but be unable to represent this as a multiplication number sentence.<br>• May count the dots in ones or skip count incorrectly. |
| | 4d | $9 \times 5 = 45$ or<br>$5 \times 9 = 45$<br><br>An array drawn to show either of the above | **On track**<br>• Uses columns and rows to create an array that represents the question.<br>• Can write a number sentence that matches the array created.<br>• Solves the problem by skip counting in fives or by recalling a known fact.<br><br>**Further learning required**<br>• May be unable to create an array to represent the problem.<br>• May be able to solve the problem by (skip) counting but has difficulty relating this to multiplication. |
| | 4e | $6 + 6 = 12$<br>and<br>$6 \times 2 = 12$ | **On track**<br>• Understands that a double can be shown as an addition and as a multiplication.<br><br>**Further learning required**<br>• May not know what 'double' means.<br>• May not understand the link between doubles and multiplication (child may be able to write an addition but not a multiplication). |

| Topic – benchmarks / Es & Os | Question | Answer | Notes |
|---|---|---|---|
| | 4f | 30 | **On track**<br>• Solves the problem using knowledge of doubling, place value and known facts.<br>• The pupil has recall of known doubles and uses this to help solve the problem, e.g. double 10 is 20.<br>• They use partitioning to split 15 into 10 and 5, double these values and add to find the total.<br>**Further learning required**<br>• May not know what doubles means.<br>• May interpret 'double' as meaning '2 more' rather than twice as many. |
| | 4g<br>4h | 7 chews each<br>4 pieces of paper to each table | **On track**<br>• Understands what it means to 'share equally' and uses concrete materials, drawings and/or known facts to solve these problems.<br>**Further learning required**<br>• May miscount.<br>• May not share the items equally. |
| | 4i | 6 boxes | **On track**<br>• Represents and solves the problem by drawing six groups of six.<br>• Appreciates the answer is 6 **boxes** (a discussion may need to be held with the pupil to determine this).<br>**Further learning required**<br>• May have difficulty representing the problem in pictures.<br>• May form unequal groups. |
| | 4j | 9 dishes<br><br>Number sentence written should reflect the pupils thinking.<br>$5 + 5 + 5 + 5 + 5 + 5 + 5 + 5 + 5 = 45$ (so 9 fives) OR $5 \times 9 = 45$ OR $45 \div 5 = 9$ | **On track**<br>• Interprets the problem correctly, e.g. as 'How many groups of 5?' and solves it using one or more of the following ways:<br>  – Represents the carrots with concrete materials, grouping them in fives and counting the groups<br>  – Draws the carrots, groups them in fives then counts the groups |

| Topic – benchmarks / Es & Os | Question | Answer | Notes |
|---|---|---|---|
| | | | – Skip counts in 5s on a number line and counts the 'jumps'. |
| | | | – Uses a known fact. |
| | | | • Writes a number sentence that reflects their thinking. |
| | | | **Further learning required** |
| | | | • May misinterpret the problem and represent it incorrectly with concrete materials or pictures. |
| | | | • May be unable to skip count and have to count in ones. |
| | | | • May be unable to write a number sentence to match how they solved the problem. |
| | 4k | 8 ten pound notes<br><br>Number sentence written should reflect the pupils thinking.<br>10 + 10 + 10 + 10 + 10 + 10 + 10 + 10 = 80 (so 8 tens)<br>OR 10 x **8** = 80<br>OR 80 ÷ 10 = **8** | **On track**<br>• Interprets the problem correctly, e.g. as 'How many tens?'and solves it using one or more of the following ways:<br>– representing the £10 notes with concrete materials or drawings<br>– skip counting in tens, mentally ('double counting' on their fingers) or by drawing a number line.<br>– using a known fact.<br>• Writes a number sentence that reflects their thinking.<br><br>**Further learning required**<br>• May misinterpret the problem and represent it incorrectly with concrete materials or pictures.<br>• May be unable to skip count and have to count in ones.<br>• May be unable to write a number sentence to match how they solved the problem. |
| | 4l | 8 tens<br>80 ÷ 10 = 8 or<br>80 ÷ 8 = 10 | **On track**<br>• Understands the question, e.g. knows they need to count the rows on the abacus to answer the first part.<br>• May skip count in tens to find the total number of beads.<br>• May use knowledge of the ten times table to find the total number of beads (10 x 8 = 80).<br>• Correctly writes a division number sentence to match the image. |

| Topic – benchmarks / Es & Os | Question | Answer | Notes |
|---|---|---|---|
| | | | **Further learning required**<br>• May not understand the first part of the question and answer '80'.<br>• May not use skip counting or knowledge of the ten times table and count by ones to find the total number of beads.<br>• May be able to write a multiplication number sentence to match the abacus picture but not a division number sentence because they do not yet fully understand the inverse relationship. |
| **Fractions, decimal fractions and percentages**<br>MNU 1-07a<br>MNU 1-07b<br>MTH 1-07c | 5a | A, D | **On track**<br>• Understands that for parts to be equal they must be exactly the same size.<br>**Further learning required**<br>• May not understand what equal means. |
| | 5b | $\frac{2}{8}$ or $\frac{1}{4}$<br>The pupil may write the answer in words | **On track**<br>• Understands how to write a fraction, in words or using fraction notation, and correctly identifies the fraction shaded.<br>**Further learning required**<br>• May confuse the numerator and denominator, e.g. writes $\frac{1}{4}$ as $\frac{4}{1}$.<br>• May record only the numerator, e.g. answers '2'. |
| | 5c | $\frac{1}{2}$, $\frac{6}{8}$ | **On track**<br>• Identifies fractions of a single item correctly.<br>• Understands what the numerator and denominator are and uses these correctly to write each fraction.<br>**Further learning required**<br>• May not understand what the numerator and denominator are and therefore make mistakes when writing fractions. |
| | 5d | $\frac{3}{6}$, 3 bars coloured<br>$\frac{1}{3}$, 2 bars coloured<br>$\frac{2}{5}$, 2 bars coloured<br>$\frac{3}{4}$, 6 bars coloured | **On track**<br>• Can represent parts of a whole correctly.<br>• Correctly uses the diagram to write each fraction with the correct numerator and denominator.<br>**Further learning required**<br>• May confuse the numerator and denominator.<br>• May record only the numerator or denominator correctly. |

| Topic – benchmarks / Es & Os | Question | Answer | Notes |
|---|---|---|---|
| | 5e | $\frac{3}{4}$<br><br>May show two diagrams shaded as $\frac{3}{4}$ and $\frac{3}{5}$<br><br> | **On track**<br>• Correctly identifies the larger fraction in each example and can draw diagrams to prove their answers.<br>• May prefer to justify their answers orally, for example, explains that $\frac{3}{4}$ is larger than $\frac{3}{5}$ because the fewer parts the same whole is split into the larger each part will be. |
| | 5f | $\frac{4}{6}$<br><br>May show two diagrams shaded as $\frac{4}{6}$ and $\frac{3}{10}$<br><br> | **Further learning required**<br>• May identify the larger fraction in each case but be unable to justify their answers diagramatically or orally.<br>• May be able to compare three fifths and three quarters but find comparing three tenths and four sixths more challenging. |
| | 5g | Number line drawn with the following fractions marked:<br>$4, 4\frac{1}{2}, 5, 5\frac{1}{2}, 6, 6\frac{1}{2},$<br>$7, 7\frac{1}{2}, 8, 8\frac{1}{2}, 9, 9\frac{1}{2}, 10$ | **On track**<br>• Knows that two halves equal one whole.<br>• Counts in halves from 4 to 10 on a number line.<br>**Further learning required**<br>• May struggle to count in halves or may revert to ones midway through. |
| | 5h<br>5i<br>5j | £9<br>7 pencils<br>12 seeds | **On track**<br>• Connects finding a fraction of a quantity with the concept of division.<br>• Represents and solves each problem using concrete materials, drawings, diagrams or known facts.<br><br>**Further learning required**<br>• May not link finding a fraction of a quantity with division and so struggle to interpret the problem.<br>• May be able to show $\frac{1}{5}, \frac{1}{4}$ and/or $\frac{1}{3}$ of an object but have difficulty understanding that $\frac{1}{5}, \frac{1}{4}$ and $\frac{1}{3}$ of a quantity can be more than one item. |
| **Money**<br>MNU 1-09a<br>MNU 1-09b | 6a<br>6b<br>6c<br>6d<br>6e<br>6f | £1.60<br>£6.50<br>£6. 88<br>570p<br>333p<br>879p | **On track**<br>• Can convert between pounds and pence, e.g. knows that a £1 coin, a 50p coin and a 2p coin have a total value of £1.52 or 152p.<br>• Can add collections of notes and coins (total value less than £10) and record these accurately using money notation (pounds and pence). |

| Topic – benchmarks / Es & Os | Question | Answer | Notes |
|---|---|---|---|
| | | | **Further learning required**<br>• May not know that £1 equals 100p.<br>• May be able to work out, for example, that £1 + 50p + 10p is 'one pound sixty' but is unable to equate this with 160p.<br>• May record amounts in pounds and pence incorrectly, e.g. writes £5.70 as £5.70p or £5.7. |
| | 6g | 73p | **On track**<br>• Identifies addition as the correct operation and can add 50 + 23 mentally to solve the problem.<br><br>**Further learning required**<br>• May not understand the problem.<br>• May rely on a written method to calculate 50 + 23. |
| | 6h | 45p + 36p = 81p<br>81p – 65p = 16p | **On track**<br>• Understands that the problem is in two parts and therefore requires two calculations.<br>• Chooses an appropriate method (for example, a number line or place value partitioning) and adds to find the total cost of the items Finlay wants to buy.<br>• Uses counting on or counting back to calculate the difference between how much Finlay has and how much he needs.<br><br>**Further learning required**<br>• May struggle to interpret the problem or is only able to complete part of it.<br>• May choose the wrong operation, e.g. adds 65p + 45p + 36p.<br>• May make calculation errors. |
| | 6i<br>6j<br>6k | 35p<br>52p<br>75p | **On track**<br>• Uses counting on or counting back to calculate change.<br><br>**Further learning required**<br>• May choose the wrong operation, e.g. adds instead of subtracting.<br>• May count on or back but does so incorrectly, e.g. includes the starting number when jumping on a number line.<br>• May attempt to solve the problem by partitioning, e.g. thinks '60 + 40 = 100 and 5 + 5 = 10 and answers 45p in question 6i. |

| Topic – benchmarks / Es & Os | Question | Answer | Notes |
|---|---|---|---|
| **Time**<br>MNU 1-10a<br>MNU 1-10b<br>MNU 1-10c | 7a | 7 minutes | **On track**<br>• Understands the problem and solves it by counting on (8 + ? = 15) or subtracting (15 − 8 = ?).<br><br>**Further learning required**<br>• May misinterpret the problem, e.g. uses the number four and/or three<br>• May choose the wrong operation and/or make calculation errors |
| | 7b | Amman<br>24 hours in one day | **On track**<br>• Knows how many hours are in a day and uses this information to compare duration.<br><br>**Further learning required**<br>• May be unable to interpret the problem.<br>• May not know how many hours are in a day. |
| | 7c | 98 − 74 = 24<br>24 more months | **On track**<br>• Understands the problem and uses a suitable strategy to find the difference, e.g. counts on or back on an empty number line or subtracts using place value partitioning.<br><br>**Further learning required**<br>• May be unable to interpret the problem.<br>• May calculate inaccurately. |
| | 7d<br>7e | 7 days<br>29th July | **On track**<br>• Can calculate duration of time (days), aided by a calendar if necessary.<br>• Can read a simple calendar.<br><br>**Further learning required**<br>• May be unable to use a calendar to calculate the duration of events. |
| | 7f | 12 | **On track**<br>• Knows there are 12 months in the year.<br><br>**Further learning required**<br>• May not know the months of the year. |
| | 7g | 48 | **On track**<br>• Knows there are 12 months in the year.<br>• Can select and use an appropriate method to find the number of months in four years, e.g. repeated addition (12 + 12 + 12 + 12) or place value partitioning (10 x 4) + (2 x 4).<br><br>**Further learning required**<br>• May not know there are 12 months in the year.<br>• May be unable to calculate the number of months in four years or calculates inaccurately. |

| Topic – benchmarks / Es & Os | Question | Answer | Notes |
|---|---|---|---|
| **Measurement**<br>MNU 1-11a<br>MNU 1-11b | 8a | Pupil draws lines measuring 4 cm and 18 cm (to within +/– 3 mm) | **On track**<br>• Understands that 'cm' means centimetres.<br>• Uses a ruler to accurately draw lines of specified lengths in centimetres.<br><br>**Further learning required:**<br>• The pupil may not start at zero, creating a line that is too long or too short.<br>• The pupil may struggle to keep the line straight. |
| | 8b<br><br><br>8c | Answer depends on size of paper and sticky notes<br><br>Answer depends on size of desk child is working at and the size of the pieces of paper used | **On track**<br>• Can estimate and measure area in non-standard units (sticky notes and paper).<br><br>**Further learning required**<br>• Does not have an estimation strategy.<br>• Does not place sticky notes or paper in a regular way. |
| | 8d | Millilitres<br>Litres | **On track**<br>• Child's explanations of their answers show an awareness of the relative size of litres and millilitres.<br><br>**Further learning required**<br>• May be unfamiliar with standard units of measurement (capacity). |
| | 8e | 3 metres 24 centimetres<br>5 metres 1 centimetre | **On track**<br>• Uses standard units of measurement (cm /m) to read and record height accurately.<br>• Understands that there are 100 cm in 1 m.<br>• Converts correctly between cm and m.<br><br>**Further learning required**<br>• May be unfamiliar with standard units of measurement (length/height).<br>• May not know there are 100 centimetres in 1 metre, so be unable to convert between these units, e.g. records 324 cm as 32 m 4 cm.<br>• May record answer using no units or the incorrect units, for example records 501 cm as 5 cm 1 m. |

| Topic – benchmarks / Es & Os | Question | Answer | Notes |
|---|---|---|---|
| | 8f<br>8g | 2 kg 605 g<br>8 kg 9 g | **On track**<br>• Uses standard units of measurement (kg /g) to record mass accurately.<br><br>**Further learning required**<br>• May be unable to distinguish between kg and g.<br>• May record answer using no units or the incorrect units. |
| | 8h<br>8i | 7 cm<br>11 cm | **On track**<br>• Reads the scale accurately and records each length using the word centimetres or the abbreviation cm.<br><br>**Further learning required**<br>• May be unable to read a scale or record standard units correctly.<br>• May record answer using no units or the incorrect units. |
| | 8j<br>8k | 4 kg<br>34 kg | **On track**<br>• Reads each scale accurately to give the mass of each object.<br>• Understands what the graduations represent on each scale.<br><br>**Further learning required**<br>• May be unable to read a scale or standard units correctly.<br>• May be able to read the first scale but have difficulty reading the second scale where the value for each interval must be identified.<br>• May record answers using no units or the incorrect unit, e.g. records the mass of the person as 34 g. |
| | 8l | 400 ml | **On track**<br>• Can work out the value of each interval on the scale (50 ml) and record the volume of water in milliliters (ml).<br><br>**Further learning required**<br>• May be unable to read a scale or standard units correctly.<br>• May record answer using no units or the incorrect units. |

| Topic – benchmarks / Es & Os | Question | Answer | Notes |
|---|---|---|---|
| **Mathematics, its impact on the world, past, present, future**<br><br>MTH 1-12a | 9a<br><br><br><br>9b | Various answers are possible, for example: Nurses – measure out medicine, to measure and record temperature. Baker – measure out ingredients, use timers, give change to customers. | **On track**<br>• Recognises different ways in which numbers are used in the world of work.<br><br>**Further learning required**<br>• May be unfamiliar with how numbers are used in everyday life. |
| **Patterns and relationships**<br><br>MTH 1-13a<br>MTH 1-13b | 10a | | **On track**<br>• Can identify and continue the pattern.<br><br>**Further learning required**<br>• May not understand that patterns can be made with shapes.<br>• May not understand that a pattern can include consecutive, identical shapes. |
|  | 10b | 15, 18, 21<br>60, 80 | **On track**<br>• Identifies and continues/completes each number sequence accurately, skip counting in 3s and in 10s.<br><br>**Further learning required**<br>• May not understand that there is a rule to follow, in order to complete a sequence. |
| **Expressions and equations**<br><br>MTH 1-15a<br>MTH 1-15b | 11a<br>11b<br>11c | 15<br>25<br>9 | **On track**<br>• Understands that the equals sign symbolised balance in a number sentence.<br>• Solves the equations using known facts or by counting on/counting back on a number line.<br>• May use the commutative law to simplify the calculation, e.g. changes $? + 16 = 25$ to $16 + ? = 25$.<br>• May use the inverse relationship, e.g. solves $50 - ? = 25$ by thinking $25 + ? = 50$.<br><br>**Further learning required**<br>• May be unable to relate the equations to 'part-part-whole' and so have difficulty choosing a method to solve them.<br>• May misinterpret and/or misrepresent the equations, e.g. adds the two known numbers to find the unknown in question a and c. |

| Topic – benchmarks / Es & Os | Question | Answer | Notes |
|---|---|---|---|
| **2D shapes and 3D objects**<br>MTH 1-16a<br>MTH 1-16b | 12a | Four corners – square, rectangle<br><br>More than four sides – hexagon, pentagon<br><br>One straight side – semi-circle<br><br>Fewer than four sides – right-angled triangle, circle, semi-circle<br><br>Curved sides – circle, semi-circle<br><br>Only one right angle – right-angled triangle | **On track**<br>• Correctly identifies 2D shapes and their properties, using appropriate language, e.g. corners, sides and curved.<br><br>**Further learning required**<br>• May be unfamiliar with language relating to the task. |
| | 12b | Triangular prism<br><br>Cube or cuboid | **On track**<br>• Correctly identifies 3D objects and their properties, using appropriate language, e.g. vertices, edges.<br><br>**Further learning required**<br>• May be able to identify objects with the properties asked for but be unable to name them.<br>• May be unfamiliar with the words vertices and edges. |
| | 12c | Cuboid | **On track**<br>• Can identify 3D objects (in this case a cuboid) from 2D drawings.<br><br>**Further learning required**<br>• May be able to identify a cuboid from a collection of 3D objects but not from a 2D image. |
| | 12d | Answers might include: triangle, square, hexagon, rectangle | **On track**<br>• Understands what tiling is and the shapes that can be used to do this.<br><br>**Further learning required**<br>• May not realise that tiling can incorporate two different shapes.<br>• May not understand that tiles need to meet on all sides. |

| Topic – benchmarks / Es & Os | Question | Answer | Notes |
|---|---|---|---|
| **Angles, symmetry and transformation** <br> MTH 1-17a <br> MTH 1-18a <br> MTH 1-19a | 13a | D | **On track** <br> • Understands what a quarter turn is and can choose a suitable method to solve the problem, e.g. by turning the paper through 90 degrees. <br> **Further learning required** <br> • May not know what a quarter turn is. <br> • May have difficulty visualising the same shape from different perspectives. |
| | 13b | Triangle, square and rectangle ticked. | **On track** <br> • Knows what a right angle is and correctly identifies the shapes with at least one right angle. <br> • May use a right angle 'measure' such as the corner of a jotter. <br> **Further learning required** <br> • May not know what a right angle. <br> • May make errors when checking for right angles. |
| | 13c | Square A1 <br> Circle C3 <br> Hexagon E5 <br> Pentagon F2 <br> Star B4 | **On track** <br> • Successfully uses grid references to identify where shapes are. <br> • Appreciates horizontal then vertical, i.e. letter then number. <br> **Further learning required** <br> • Does not know what grid reference are. <br> • May write the grid reference the wrong way round. <br> • May confuse the hexagon and pentagon. |
| | 13d | Symmetrical: heart, square, isosceles triangle, pentagon ticked <br><br> Not symmetrical: 6-sided irregular shape and scalene triangle crossed | **On track** <br> • Knows what symmetrical means. <br> • Correctly identifies the symmetrical shapes. <br> • May use a mirror to check. <br> **Further learning required** <br> • May not know what symmetrical means. <br> • May be unable to choose a method for checking and rely on 'guess work'. |

| Topic – benchmarks / Es & Os | Question | Answer | Notes |
|---|---|---|---|
| **Data handling and analysis** MNU 1-20a | 14a | Football 6 5 Car | **On track** • Understands how to read grouped tallies and is able to extract the information requested from the table. • Understands the phrase most/least popular. **Further learning required** • Unable to interpret the problem, e.g. is unfamiliar with the vocabulary used. • Miscounts the tally marks. |
| | 14b | 5 faces in hazel bar 8 faces in blue bar 3 faces in green bar 9 faces in brown bar | **On track** • Correctly interprets the information given and uses this to complete the bar graph correctly. **Further learning required** • May misread the data and create a picture graph that looks right but has the wrong values. |
| | 14c | i.   Car ii.  Walk and bus iii. Scooter iv.  Bicycle v.   5 | **On track** • Extracts information from the graph correctly to answer the questions. **Further learning required** • May not understand the language of most and least popular. • May be unable to choose an appropriate strategy for questions iv and v. |
| **Ideas of chance and uncertainty** MNU 1-22a | 15 | Certain Unlikely Unlikely Impossible | **On track** • Understands the vocabulary of probability and can use it to make judgements about the likelihood of events occurring. **Further learning required** • May not understand the mathematical interpretation of the 'chance words', e.g. refusing to use 'impossible' in the belief that 'nothing is impossible'. |

# Yearly progress check 1C

## Resources

**Resources provided online**

Square and triangle template for tiling for use with 12c (see Resource 1)

**Resources needed and available in the setting**

Ruler  •  Squared paper

## Questions

1.  a   Is 491 closer to 400 or 500? Write 491 in the correct place on the number line.

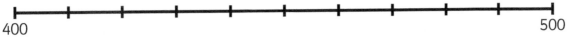

400                                                                        500

b   Is 858 closer to 800 or 900? Write 858 in the correct place on the number line.

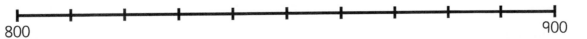

800                                                                        900

c   Finlay's mum and dad are looking at the price of a holiday to Spain. It costs £389 per adult and £234 per child.

Estimate how much will it cost for 2 adults and 2 children by rounding to the nearest ten.

_____

Show how you worked it out.

2. **579**

   a  Write this number in words. _____

   Now reverse the digits and write this new number in words. _____

   _____

   b  Write the number that is 5 hundreds, 4 tens and 8 ones using digits.

   _____

   Now write this number in words.

   _____

   c  Write the number shown by this place value house in numerals and in words.

    _____

   _____

   d  Write the number shown by this place value house in numerals and in words.

    _____

   _____

   e  Count **backwards** from 801.

   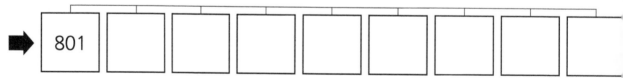

   f  Write the number that is 10 more than and 100 more than:

   701 _____   _____

   398 _____   _____

# Yearly progress check 1C

g   Write the number that is 10 less than and 100 less than:

   i   600 _____   _____

   ii  209 _____   _____

h   There are 20 pencils in each pack. Skip count to find the total number of pencils.

_____

i   Complete the number sequence.

j   For each number, write the value of the underlined digit.

650 _____

347 _____

999 _____

405 _____

k  Write the number represented by the base 10 blocks.

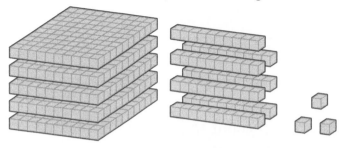

Partition the number into hundreds, tens and ones in two ways.

l  Find the missing number. The number at the top of each triangle is the sum of the two bottom ones.

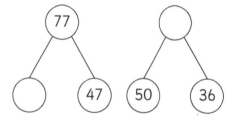

m **897**

Write a three-digit number, in numerals and words, that is greater than this number.

_____

_____

Write a three-digit number, in numerals and words, that is less than this number.

_____

_____

n  Write the numbers in the correct order from largest to smallest.

674, 601, 67, 60, 650, 611, 699

_____

# Yearly progress check 1C

o Isla was looking at the places runners had come in a race. Write these ordinal numbers in words.

Lucy **45**th _____

Lola **106**th _____

Mike **21**st _____

Tony **198**th _____

3. a Solve these number sentences:

5 + 7 = _____

175 + 7 = _____

15 − 4 = _____

515 − 4 = _____

b Finlay has 463 marbles. Isla gives him 10 more.

How many does he have now? _____

c Amman had 375 stickers. He gives Nuria 100.

How many does he have now? _____

d Solve these number sentences:

543 + 300 = _____

850 − 500 = _____

e Write a number sentence for this number line and solve it.

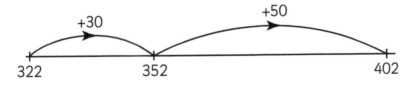

# Yearly progress check 1C

f   Draw an empty number line to help you complete this number sentence.

34 + [          ] = 100

g   Write a fact family for the bar model.

| 100 | |
| 74 | 26 |

_____   _____

_____   _____

h   Double these numbers. Explain how you worked them out.

69 _____

85 _____

i   Halve these numbers. Explain how you worked them out.

76 _____

106 _____

j   Isla has different scores for her maths quizzes: 17, 18, 42

What is her total score? _____

Show how you worked it out.

k  Finlay counted how many cars he saw on his way to school each day for one week.

| Monday | Tuesday | Wednesday | Thursday | Friday |
|--------|---------|-----------|----------|--------|
| 13     | 7       | 21        | 16       | 9      |

How many cars did Finlay see in one week of school?

_____

Show how you solved the problem.

l  Complete these number sentences.

$91 -$ ☐ $= 82$

☐ $- 8 = 66$

m Find the answer to each subtraction using a strategy of your choice. Use the same numbers to write an addition partner for each subtraction.

$55 - 43 =$ _____     ☐ $+$ ☐ $=$ _____

$82 - 38 =$ _____     ☐ $+$ ☐ $=$ _____

Solve these word problems using a method of your choice. Show how you worked the answers out.

n  365 people arrive at the cinema to see a new film. Eight people do not get a seat. How many seats are there in the cinema?

o  Isla has double the number of stickers that Nuria has. If Nuria has 68 stickers how many does Isla have?

p  There were 189 passengers on a train. Some passengers got off at the next station and 174 were left on the train. How many people got off at the station?

q  Amman's dad was making a shed. He had some screws and 167 nails. There were 189 screws and nails altogether. How many screws were there?

r  Choose a strategy to solve each number sentence. Show how you worked each answer out.

324 + 65 = _____

89 + 234 = _____

548 − 79 = _____

725 − 95 = _____

270 + [          ] = 815

657 − 254 = [          ]

[          ] + 287 = 368

s Complete a Think Board for each question

Amman has 356 toy soldiers. Finlay has 234 more than Amman. How many toy soldiers does Finlay have?

| Bar model | Number sentence and answer |
| --- | --- |
| | |
| **Number line** | |

The postman delivers 246 letters. When he stops for lunch he still has 168 to deliver. How many letters did he start with?

| Bar model | Number sentence and answer |
| --- | --- |
| | |
| **Number line** | |

Isla's class had a box of glue sticks. The class used 145 of them and by the end of the year there were 231 left. How many glue sticks were there in the box to begin with?

| Bar model | Number sentence and answer |
|---|---|
| | |
| **Number line** | |

t i P4 are raising money for charity. Their sponsored read raises £456 and their book sale raises £129. How much money do they still need to reach their target of £850?

ii Nuria is baking a cake. All of the ingredients together weigh 950 g. She needs 400 g of flour, 205 g sugar, 80 g of eggs and some chocolate. How much chocolate does Nuria need?

4. a Isla has 27 marbles and shares them equally between herself, Finlay and Amman. How many do they get each?

b Logan was sharing out sweets into her party bags. She had 27 sweets to share equally between 5 children. Is this possible? Explain your thinking.

_____

_____

How many sweets were in each bag? _____

How many sweets were left over? _____

c Answer the following questions. You may use jottings to help you:

$19 \div 2 =$ _____

$28 \div 3 =$ _____

d There are 8 classes in the school. Each class has three finalists for the poetry competition. How many finalists are there?

Show how you worked this out. _____

e Draw an array for this multiplication and solve it.

$7 \times 6 =$ _____

f Mrs Clark has been collecting plant seeds. She has collected nine packs of 100 seeds. How many seeds does she have?

_____

g Answer the following questions. Explain your thinking:

300 ÷ 100 = _____

500 ÷ 100 = _____

h Write a multiplication partner for each division:

35 ÷ 5 = 7 _____

18 ÷ 2 = 9 _____

i Write two multiplication facts for the first triangle and two division facts for the second triangle.

  _____    _____

  _____    _____

j What number is missing from this multiplication and division triangle? Explain your thinking.

  _____

k Work out the answer to this division. Show how you did it. $600 \div 2 =$

_____

l There are 200 children going on the school trip. Each minibus can only take 20 children.

How many minibuses will the school need to take all of the children?

Show how you worked this out. _____

m Amman is saving up for a new tablet costing £240. He has saved £86 already.

He earns £5 a week from his paper round which he does for 9 weeks.

How much more does Amman now need to buy his tablet?

Show how you worked out each part of the problem. _____

5. a  Draw a bar model to solve the following:

one third of £27

one quarter of 44 glue sticks

b  The school football team has scored 54 goals this season. Finlay has scored one sixth of them. How many goals has Finlay scored? _____

c  Olivia had 40 children at her party. One fifth of them were boys. How many boys were at the party? _____

d  Write these fractions in order from largest to smallest. Use the fraction wall to help you.

$$\frac{1}{5} \qquad \frac{1}{6} \qquad \frac{1}{10} \qquad \frac{1}{2} \qquad \frac{1}{4} \qquad \frac{1}{8}$$

_____

| one tenth | one tenth | one tenth | one tenth | one tenth | one tenth | one tenth | one tenth | one tenth | one tenth |
|---|---|---|---|---|---|---|---|---|---|
| one sixth || one sixth || one sixth || one sixth || one sixth || one sixth ||
| one half |||||| one half ||||||
| one eighth || one eighth || one eighth || one eighth || one eighth || one eighth || one eighth || one eighth ||
| one quarter ||| one quarter ||| one quarter ||| one quarter |||
| one fifth || one fifth || one fifth || one fifth || one fifth ||

# Yearly progress check 1C

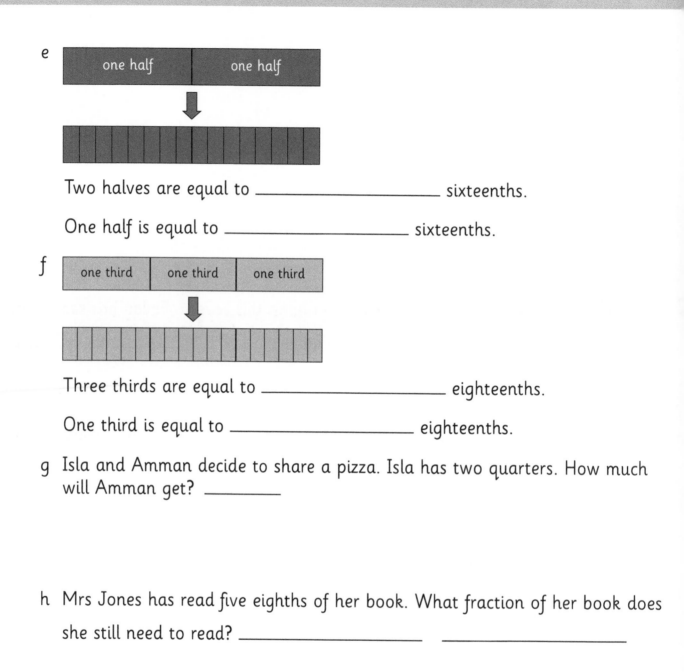

Two halves are equal to _____ sixteenths.

One half is equal to _____ sixteenths.

Three thirds are equal to _____ eighteenths.

One third is equal to _____ eighteenths.

g Isla and Amman decide to share a pizza. Isla has two quarters. How much will Amman get? _____

h Mrs Jones has read five eighths of her book. What fraction of her book does she still need to read? _____   _____

6. a  Write these amounts in £ and p

1876p _____

690p _____

1508p _____

Now make each amount using the least number of notes and coins possible.

1876p

690p

1508p

b  Tony has £9.50 in his bank account. He wants to buy some music online using his PayPal account.

Does he have enough money to pay for three songs costing £2.90 each?

_____

Show your thinking.

c Olivia and her mum are shopping. They buy bread, milk and eggs.

Olivia's mum wants to pay by debit card. She has £3 in the bank.

Does she have enough money? _____

How much more money does she need? _____

£1.10    65p    £2.25    £1.35

d Each child has some cash:

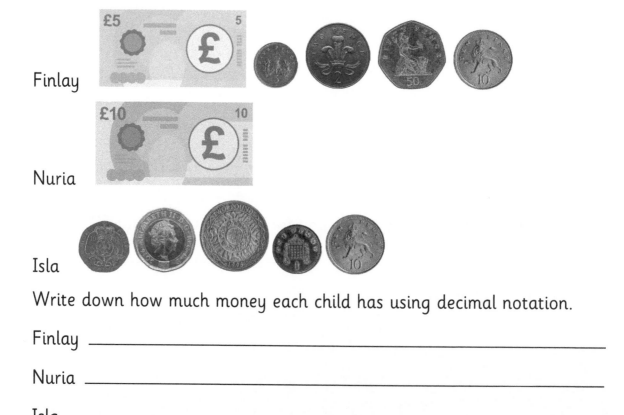

Finlay

Nuria

Isla

Write down how much money each child has using decimal notation.

Finlay _____

Nuria _____

Isla _____

How much money do Finlay and Isla have in total?

_____

How much money do all three children have altogether? Draw this amount using the least number of notes and coins.

e Mrs Clark buys some new colouring pens for the class. The pens cost £6.85. She pays with a £10 note. How much change does she get?

_____

7.

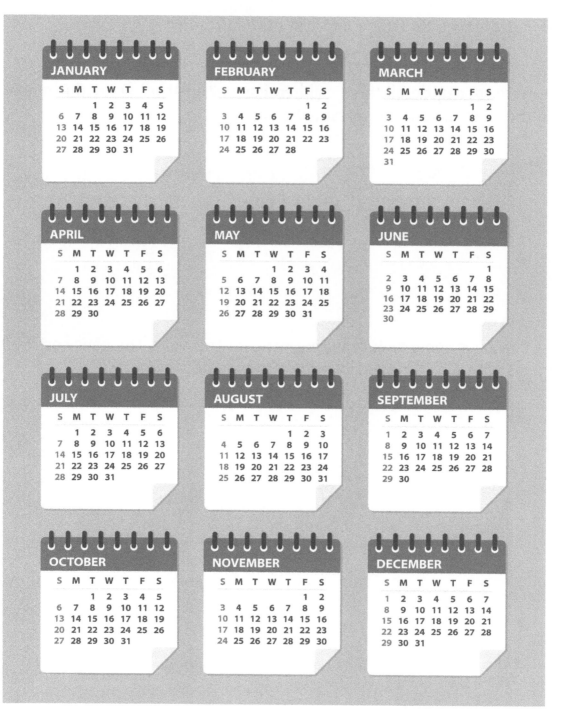

a Look at the calendar above for questions.

i What day is the 4th August? _____

ii What is the date on the second Wednesday in November? _____

iii Write the date of the second Thursday in February using words and numbers. _____

# Yearly progress check 1C

b  What could you use to:

   i   Time a race on Sports Day? _____

   ii  Make sure you got up on time in the morning? _____

   iii Check what day your friend's birthday was on? _____

   iv Tell the time with? _____

c  Draw the hands on the clocks to show;

        6:00              12:30

        5:45              8:15

d  Write each time shown here in digital form.

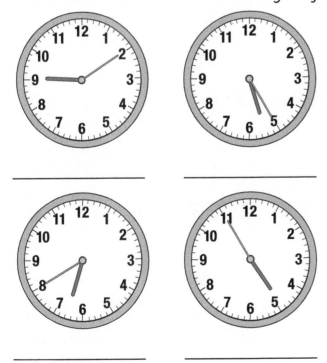

_____      _____

_____      _____

e  When do these things usually happen? Choose a.m., p.m. or both:

Get up to go to school _____

Brush your teeth _____

Eat lunch _____

Go to bed _____

f

| Leckie Channel | City 7 TV |
| --- | --- |
| 17:00 – Bigville | 17:30 – News |
| 18:00 – Cartoons | 18:00 – Weather |
| 18:30 – Peeta Planet | 18:15 – Basil Bay |
| 19:00 – News | 18:45 – Loveheart High |
| 19:30 – Weather | 19:00 – Top Cook |

What time is the weather on Leckie Channel?_____

How long is Peeta Planet on for? _____

How long is Basil Bay on for?     _____

How long is it between the start time of Bigville and the end time of the News on the Leckie Channel?

_____

8. a  You will need a ruler:

Draw a line 5 centimetres long.


Draw a line nine and a half centimetres long.


b  What is the length of the bar?

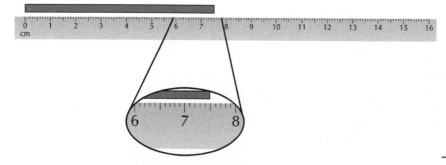

c What is the length of this bar in millimetres?

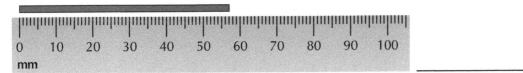

_____

d What is the mass of the apples? State your answer in kilograms.

_____

e How heavy are the carrots? State your answer in kilograms.

_____

f How much water is in the jug? State your answer in both litres and millilitres.

_____    _____

g What is the area of this shape in squares?

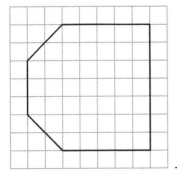

_____

h Draw two different shapes each with an area of 12 squares.

Draw a shape with an area of 16 ½ squares.

i Write a number sentence to show how many squares would fit into this rectangle.

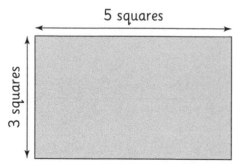

# Yearly progress check 1C

9. Here is an image of the Chinese Mandarin number system. Work out what each symbol means.

| 一 | 二 | 三 | 四 | 五 | 六 | 七 | 八 | 九 | 十 |
|---|---|---|---|---|---|---|---|---|---|
| 十一 | 十二 | 十三 | 十四 | 十五 | 十六 | 十七 | 十八 | 十九 | 二十 |
| 二十一 | 二十二 | 二十三 | 二十四 | 二十五 | 二十六 | 二十七 | 二十八 | 二十九 | 三十 |

Write the number 28 in Chinese Mandarin. _____

Write the number 16 in Chinese Mandarin. _____

10. a Write the missing numbers in each of these number patterns.

3, 6, [    ], [    ], 15, [    ]

18, 12, 6, [    ]

b Continue each of these number patterns.

4, 8, 12, 16, [    ], [    ], [    ]

45, 60, 75, [    ], [    ], [    ]

c Using the rule and the starting numbers provided, write the next five numbers for each pattern.

| Rule | Starting number | 1st term | 2nd term | 3rd term | 4th term | 5th term |
|---|---|---|---|---|---|---|
| Add 6 | 4 | | | | | |
| Subtract 20 | 200 | | | | | |
| Add 12 | 15 | | | | | |

d For the table below, fill in the gaps and decide what the rule is

| 1st term | 2nd term | 3rd term | 4th term | 5th term | Rule |
|---|---|---|---|---|---|
| 45 | | | 30 | 25 | |
| 26 | 29 | 32 | | 38 | |

11. Solve the following equations.

    a  $7 + \boxed{\phantom{xxxxx}} = 14$

    b  $\boxed{\phantom{xxxxx}} = 6 \times 4$

    c  $\boxed{\phantom{xxxxx}} - 9 = 11$

    d  $15 \div \boxed{\phantom{xxxxx}} = 3$

    e  Find the values of ⬤ , ◇ and △ .

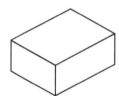

⬤ + ⬤ = 10

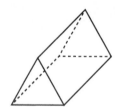

6 + △ = 10

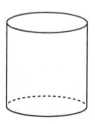

⬤ + △ = ◇

12. a  Name and describe each 3D object using the words faces and vertices.

_____ _____ _____

_____ _____ _____

_____ _____ _____

b Which 3D object is this?

| front view | side view | top view |
|---|---|---|
| △ | △ | ◯ |

_____

13. a A right angle measures ⬚ degrees.

b Tick the shapes / objects that have at least one right angle.

   A          B          C          D          E

c

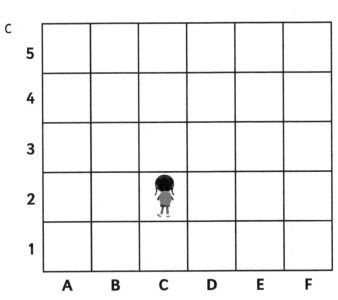

Nuria is standing in C2, Facing C4.

She follows these directions:

Forwards 2 > Right 90 > Forward 3> Left 90 > Forward 1 > Backwards 3

Where is Nuria now?

_____

d

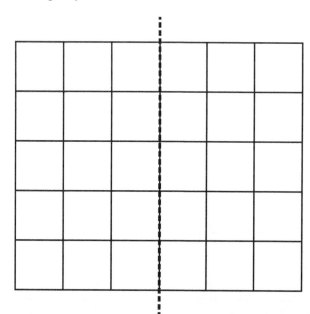

Write down directions for the following, using north, south, east and west.

i  Write directions from Finlay's house to Nuria's house.

_____

_____

ii  Write directions from Amman's house to Finlay's house.

_____

_____

e  Colour squares to create a symmetrical design using two colours. Do not colour every square.

f   Draw an object that has more than one line of symmetry.

14. a   10 children were asked, 'What month is your birthday in?'. Here are the answers:

**February**, **August**, **September**, **December**, **February**, **January**, **May**, **May**, **April**, **May**

Display this data as a table and describe one thing you notice.

b   Copy and complete the Carroll diagram using the numbers below:

1 , 2 , 3, 4 , 5, 6, 7, 8, 9, 10, 11, 12, 13, 14, 15, 16, 17, 18

|  | Odd numbers | Even numbers |
|---|---|---|
| Less than ten |  |  |
| Ten or more |  |  |

c Complete the Venn diagram using the numbers below:

5, 10, 15, 20, 25, 30, 35, 40, 45, 50, 55, 60, 70, 80, 90, 100

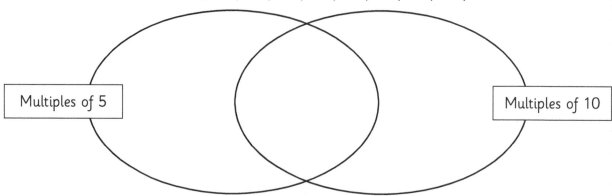

d

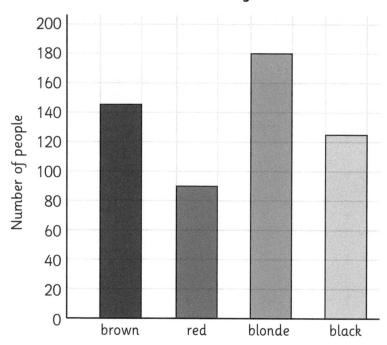

**Hair Colours in Anytown School**

Use the graph to complete the table. Put a tick in the correct box.

| Statement | True | False |
|---|---|---|
| More children have brown hair than red hair | | |
| Most children in the school have brown hair | | |
| Black is the least common hair colour in the school | | |
| There are more than 500 pupils in Anytown School | | |

e **Number of people going to the gym over a four-day period**

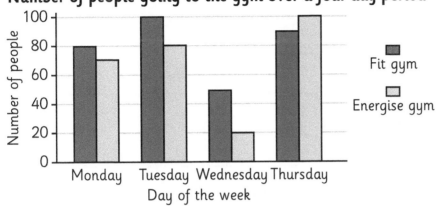

Fit gym

Energise gym

This graph shows the number of people going to two different gyms.

Look at the graph and answer the questions.

How many people went to Fit gym on Tuesday?

_____

Which day is most popular to go to the gym?

_____

How many more people went to Fit gym than Energise gym on Wednesday?

_____

f Use the data in the table to make a bar graph about favourite snacks.

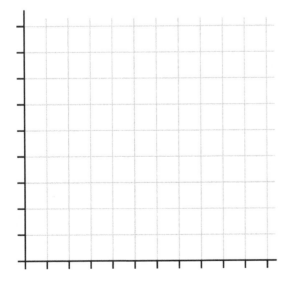

| Snack | Number of people (tally marks) |
|---|---|
| crisps | ЖЖ ЖЖ ЖЖ |
| chocolate | ЖЖ ЖЖ \|\| |
| fruit | ЖЖ \|\| |
| breadstick | ЖЖ ЖЖ ЖЖ \|\| |

15. a  Answer using **certain, likely, possible** or **impossible.** Write your answer next to each statement:

You will drink something tomorrow.

_____

You will fall in the playground.

_____

There will be a giraffe in the playground.

_____

b  State how likely the following are using **will happen**, **could happen** or **will not happen**. Write your answer next to each statement:

During the night it will be dark.

_____

It will be windy tomorrow.

_____

It will rain chocolate.

_____

You will meet someone famous next week.

_____

| Topic – benchmarks / Es & Os | Question | Answer | Notes |
|---|---|---|---|
| **Estimation and rounding** <br><br> MNU 1-01a | 1a <br><br><br><br><br><br> 1b | 500 <br><br> 491 <br> ↓ <br> 400      500 <br><br> 900 <br><br> 858 <br> ↓ <br> 800      900 | **On track** <br> • Rounds to the nearest 100 correctly, using knowledge of counting in tens. <br> • Can work out the value of each graduation on the number lines and position their answers with reasonable accuracy. <br><br> **Further learning required** <br> • May be unable to work out the value of each graduation on the number lines. <br> • May not understand what each digit in the number represents. |
| | 1c | £1240 (2 adults at approximately £390 each plus 2 children at approximately £230 each) <br><br> £1250 (if total found first then rounded) <br><br> *This problem provides an opportunity to discuss the merits of each approach.* | **On track** <br> • Can use knowledge of place value and rounding to estimate the cost for one adult and one child then carry out the correct calculations to find an approximate total cost. <br> • May use strategies such as partitioning or doubling to work out the actual total cost then round the answer. <br><br> **Further learning required** <br> • May be unable to decode the problem or fail to complete all steps in the question. <br> • May round to the nearest hundred rather than the nearest ten. <br> • May not estimate, but work out the exact answer. <br> • May not understand that the context matters when rounding. |
| **Number – order and place value** <br><br> MNU 1-02a | 2a <br><br><br><br><br><br> 2b | Five hundred and seventy-nine <br> 975 <br> Nine hundred and seventy-five <br><br> 548 <br> Five hundred and forty-eight | **On track** <br> • Correctly writes the numbers in numerals and makes a reasonable attempt at spelling the number names correctly when writing them in words. <br><br> **Further learning required** <br> • May not know the number words to use. <br> • May not understand the value of each digit. <br> • May add extra zeros when writing the numbers in numerals, e.g. 50 048 for five hundred and forty-eight. |

| Topic – benchmarks / Es & Os | Question | Answer | Notes |
|---|---|---|---|
| | 2c | 608<br>Six hundred and eight | **On track**<br>• Understands that three-digit numbers are made up of hundreds, tens and ones.<br>• Correctly uses zero as a place holder when reading and writing three-digit numbers.<br>**Further learning required**<br>• May not understand the place value of each digit.<br>• May not know the number words to use. |
| | 2d | 290<br>Two hundred and ninety | |
| | 2e | 800, 799, 798, 797, 796, 795, 794, 793 | **On track**<br>• Knows the counting sequence for three-digit numbers.<br>• Correctly adds or subtracts 10 or 100 or uses the place value pattern to continue the number sequence.<br>**Further learning required**<br>• May be unsure of the counting sequence for numbers in the hundreds or has difficulty bridging a ten or a hundred.<br>• May rely on counting in ones to find 10 more/less. |
| | 2f | 711 801<br>408 498 | |
| | 2g | 590 500<br>199 109 | |
| | 2h | 160 | **On track**<br>• Uses knowledge of place value and skip counting in 2s, to help them skip count in 20s.<br>**Further learning required**<br>• May be unable to skip count in 20s.<br>• May omit numbers in the counting sequence or only get so far, e.g. can count in 20s up to 100 but not beyond. |
| | 2i | 450, 500, 550, 600, 650 | **On track**<br>• Uses knowledge of skip counting in 5s, to help them skip count in 50s.<br>**Further learning required**<br>• May be unable to skip count in 50s or struggle to skip count backwards to find the missing numbers.<br>• May need to use another strategy to get the answer, such as adding 50 using concrete materials. |
| | 2j | 600 or 6 hundreds<br>7 or 7 ones<br>90 or 9 tens<br>0 or no/zero tens | **On track**<br>• Understands that 3-digit numbers have place values of hundreds, tens and ones, from left to right, and can identify what each column is worth in a number correctly.<br>**Further learning required**<br>• May not understand the place value of 3-digit numbers or gives the wrong values. |

| Topic – benchmarks / Es & Os | Question | Answer | Notes |
|---|---|---|---|
| | 2k | 573<br>5 hundreds, 7 tens and 3 ones<br>500 + 70 + 3<br>Other answers are possible, e.g. 500 + 73; 400 + 170 + 3 etc. | **On track**<br>• Can correctly partition a 3-digit number into hundreds, tens and ones.<br><br>**Further learning required**<br>• May be able to say the number has 5 hundreds, 7 tens and 3 ones but does not equate this with 500 + 70 + 3. |
| | 2l | 30<br>86 | **On track**<br>• Completes each part-part-whole diagram correctly.<br><br>**Further learning required**<br>• May be able to find the unknown whole when the two parts are known (second example) but find the missing part example more challenging.<br>• May choose the wrong operation, i.e. does not understand that addition is required to find the unknown whole and subtraction is required to find an unknown part.<br>• May be unfamiliar with the way the problem is presented. |
| | 2m | Pupil to write a three-digit number greater than 897 in numerals and in words<br><br>Pupil to write a three-digit number less than 897 in numerals and in words | **On track**<br>• Understands the value of 897 and uses this knowledge to correctly write a number that is greater and a number that is smaller than 897 in numerals and words (accept a reasonable attempt at spelling the numbers when writing them in words).<br><br>**Further learning required**<br>• May not understand the value of 897.<br>• May not understand the words greater and smaller.<br>• The number words used do not match the number written in numerals, e.g writes 896 as 'eight nine six' or 'eight hundred and sixty-nine'. |
| | 2n | 699, 674, 650, 611, 601, 67, 60 | **On track**<br>• Uses knowledge of place value to order the numbers correctly from largest to smallest.<br><br>**Further learning required**<br>• May confuse largest and smallest and write the solution in the reverse order. |

# Yearly progress check 1C    Marking guidance

| Topic – benchmarks / Es & Os | Question | Answer | Notes |
|---|---|---|---|
| | 2o | Forty-fifth<br>One hundred and sixth<br>Twenty-first<br>One hundred and ninety-eighth | **On track**<br>• Correctly writes ordinal numbers in words (accept a reasonable attempt at spelling the number when writing it in words).<br><br>**Further learning required**<br>• May not understand the language of ordinal numbers.<br>• May write twenty-oneth or similar instead of 'twenty-first'. |
| **Number – addition and subtraction**<br>MNU 1-03a | 3a | 12 182<br>11 511 | **On track**<br>• Uses knowledge of number bonds and place value to solve the problems.<br><br>**Further learning required**<br>• May be unable to apply knowledge of number bonds to calculations with 3-digit numbers.<br>• May not see the connection between the first and second part of the questions and need to use concrete materials or count and back on a number line. |
| | 3b<br>3c | 473<br>275 | **On track**<br>• Uses place value knowledge to add and subtract 10 and 100, e.g. knows that if they add 10 to a 3-digit number, there will be 'one more ten' so the tens digit will increase from 6 to 7.<br><br>**Further learning required**<br>• May not equate 10 with 'one ten' and/or 100 with 'one hundred' and so rely on counting by ones on a number line or using concrete materials.<br>• May not interpret the question correctly and add instead of subtracting or vice versa. |
| | 3d | 843<br>350 | **On track**<br>• Uses partitioning (e.g. 500 + 300 + 40 + 3) and/or knowledge of place value (e.g. subtract 5 hundreds) to find the correct answers.<br><br>**Further learning required**<br>• May be unable to partition.<br>• May not understand the value of the digits. |
| | 3e | 322 + 80 = 402 | **On track**<br>• Interprets the number line correctly to write and solve the addition number sentence.<br><br>**Further learning required**<br>• May be unable to interpret the number line.<br>• May know they need to add 80 to 322 (50 + 30) but need to count in tens to do so. |

| Topic – benchmarks / Es & Os | Question | Answer | Notes |
|---|---|---|---|
| | 3f | 66<br>The pupil should draw an empty number line to show how they solved the problem. | **On track**<br>• Uses a number line with partitioning to efficiently get the answer.<br>**Further learning required**<br>• May only be able to jump in 1s or 10s on a number line.<br>• May not partition 34 to simplify the calculation.<br>• May count on incorrectly or wrongly and add 34 to 100. |
| | 3g | $26 + 74 = 100$<br>$74 + 26 = 100$<br>$100 - 74 = 26$<br>$100 - 26 = 74$ | **On track**<br>• Understands the part-whole relationship and can use this knowledge to write a 'fact family' (two additions and two subtractions) to match the bar model.<br>**Further learning required**<br>• May be unfamiliar with bar models.<br>• May not understand that addition is commutative.<br>• May wrongly assume that the numbers in a subtraction are interchangeable as in an addition. |
| | 3h | 138<br>170 | **On track**<br>• Understands that when we double a number we multiply it by 2 or add it to itself.<br>• Can double each number using a suitable strategy. For example, by partitioning (double 85 is 160 + 10) or using round and adjust (double 69 is double 70 minus 2).<br>**Further learning required**<br>• May not understand what 'double' means or be unable to select a suitable strategy to solve the problem. |
| | 3i | 38<br>53 | **On track**<br>• Understands that when we halve a number we divide it by 2.<br>• Can halve each number using a suitable strategy. For example, by partitioning (half of 100 is 50; half of 6 is 3; 50 add 3 equals 53).<br>**Further learning required**<br>• May not understand what 'halve' means or be unable to select a suitable strategy to solve the problem. |

| Topic – benchmarks / Es & Os | Question | Answer | Notes |
|---|---|---|---|
| | 3j | 77<br>One possible strategy is:<br>18 + 42 = 60<br>60 + 17 = 77<br><br>66<br>One possible strategy is:<br>13 + 7 = 20 and<br>21 + 9 = 30<br>20 + 30 + 16 = 66 | **On track**<br>• Adds a string of numbers correctly by looking for multiples of 10 and/or number bonds to 100.<br><br>**Further learning required**<br>• May still require concrete materials or number lines to total a string of numbers. |
| | 3k | | |
| | 3l | 91 – **9** = 82<br>**74** – 8 = 66<br>Pupils should explain their thinking orally or in writing | **On track**<br>• Understands the inverse relationship and so can chose between counting on and counting back to solve missing subtrahend (first example) and missing minuend (second example) problems.<br><br>**Further learning required**<br>• May be able to solve the first problem using a 'count down to' strategy but wrongly believes the same strategy will work for the second problem (the child does not yet understand subtraction as the inverse of addition). |
| | 3m | 12<br>12 + 43 = 55 or<br>43 + 12 = 55<br><br>44<br>44 + 38 = 82 or<br>38 + 44 = 82 | **On track**<br>• Uses known facts, partitioning and or a number line to solve the problem.<br>• Understands the inverse relationship and can write an addition to partner each subtraction.<br><br>**Further learning required**<br>• May use a less efficient strategy such as counting in ones.<br>• May not understand the inverse relationship and so be unable to write an addition number sentence to 'match' each subtraction. |
| | 3n | 357 seats<br>Pupils should be able to show/explain how they solved the problem | **On track**<br>• Can interpret the question and identifies subtraction as the required operation.<br>• May represent the problem with concrete materials, drawings or diagrams (e.g. bar model) to help them visualise the problem.<br>• Chooses an appropriate method to solve the problem. One or a combination of the following strategies would be appropriate: known facts, partitioning, counting back on a number line, rounding and compensating. |

| Topic – benchmarks / Es & Os | Question | Answer | Notes |
|---|---|---|---|
| | | | **Further learning required**<br>• May be unable to interpret/represent the word problem.<br>• May choose the wrong operation, e.g. adds 365 + 8.<br>• May choose an inefficient strategy, e.g. tries to count on from 8 up to 365.<br>• May use one of the above strategies incorrectly or make calculation errors. |
| | 3o | 136 stickers<br>Pupils should be able to show/explain how they solved the problem | **On track**<br>• Understands what 'double' means and interprets the question correctly.<br>• May represent the problem with concrete materials, drawings or diagrams (e.g. bar model) to help them visualise double/twice as many.<br>• Uses a doubling strategy to solve the problem, e.g. partitioning $(2 \times 60) + (2 \times 8)$; $(60 + 60) + (8 + 8)$ or rounding and compensating $(2 \times 70) - 4$.<br><br>**Further learning required**<br>• May be unable to interpret/represent the word problem.<br>• May choose an inefficient strategy and/or make calculation errors. |
| | 3p | 15 passengers<br>Pupils should be able to show/explain how they solved the problem | **On track**<br>• May represent the problem with concrete materials, drawings or diagrams (e.g. bar model) to help them visualise the problem.<br>• Chooses an appropriate method to solve the problem. One or a combination of the following strategies would be appropriate: known facts, partitioning, counting on or back on a number line, rounding and compensating.<br><br>**Further learning required**<br>• May be unable to interpret/represent the word problem.<br>• May choose the wrong operation, e.g. adds 189 + 174.<br>• May choose an inefficient strategy and/or make calculation errors. |

| Topic – benchmarks / Es & Os | Question | Answer | Notes |
|---|---|---|---|
| | 3q | 22 screws<br>Pupils should be able to show/explain how they solved the problem | **On track**<br>• Can interpret the question and identifies subtraction as the required operation.<br>• May represent the problem with concrete materials, drawings or diagrams (e.g. bar model) to help them visualise the problem.<br>• Chooses an appropriate method to solve the problem. One or a combination of the following strategies would be appropriate: known facts, partitioning, counting on or back on a number line, rounding and compensating.<br>**Further learning required**<br>• May be unable to interpret/represent the word problem or does so incorrectly, e.g. answers 189 screws.<br>• May choose an inefficient strategy and/or make calculation errors. |
| | 3r | $324 + 65 = \mathbf{389}$<br>$89 + 234 = \mathbf{323}$<br>$548 - 79 = \mathbf{469}$<br>$725 - 95 = \mathbf{630}$<br>$270 + \mathbf{545} = 815$<br>$657 - 254 = \mathbf{403}$<br>$\mathbf{81} + 287 = 368$ | **On track**<br>• Understands that different strategies can be used to solve the same problem.<br>• Correctly uses one or more of the following strategies: known facts, partitioning, counting on and back on a number line, rounding and compensating.<br>**Further learning required**<br>• May use one of the above strategies incorrectly and therefore gets the incorrect answer.<br>• May not understand the inverse relationship between addition and subtraction and so is unable to solve $270 + ? = 815$ and $? + 287 = 368$, or does so incorrectly. |
| | 3s | 590 toy soldiers<br><br>414 letters<br><br>376 glue sticks | **On track**<br>• Can interpret and visualise each question in terms of 'part-part-whole'.<br>• Understands that the same problem can be represented and solved in different ways.<br>• Correctly uses one or more of the following strategies: known facts, partitioning, count on, multiples of 10, rounding and compensating.<br>• Completes a Think Board for each problem. Each Think Board should have a number sentence, bar model and number line that clearly illustrates the pupil's ability to solve non-standard addition problems and understanding of the inverse relationship. |

| Topic – benchmarks / Es & Os | Question | Answer | Notes |
|---|---|---|---|
| | | | **Further learning required**<br>• May be unable to interpret/represent the word problems and so chooses the wrong operation, e.g. subtraction instead of addition.<br>• May use one of the above strategies incorrectly and therefore gets the incorrect answer.<br>• May be unfamiliar with a Think Board. |
| | 3t | i £265<br>456 + 129 = 585<br>850 – 585 = 265<br><br>ii 265 g<br>400 + 205 + 80 = 685<br>950 – 685 = 265 | **On track**<br>• Can interpret, represent and solve two-step word problems involving addition and subtraction.<br>• Uses one or a combination of the following strategies: known facts, partitioning, counting on and back on a number line, rounding and compensating.<br><br>**Further learning required**<br>• May misinterpret the problem and/or choose the wrong operations.<br>• May only complete part of each problem.<br>• May make a calculation error on one part of the question resulting in the incorrect answer. |
| **Number – multiplication and division**<br>MNU 1-03a | 4a | 9 marbles each | **On track**<br>• Understands that 'share equally' means give the same amount to each.<br>• Can represent and solve the problem using a method of their choice, e.g. concrete materials, drawings or diagrams, skip counting in 3s or recalling a known fact, i.e. $3 \times 9 = 27$ or $27 \div 3 = 9$.<br><br>**Further learning required**<br>• May be unable to represent and solve the problem.<br>• The above methods show a progression in the way a child thinks about a sharing problem. Pupils dependent on counting in ones, using concrete materials or drawings, need practice in skip counting collections in 3s (repeated addition/subtraction) and connecting this to a number sentence. |

| Topic – benchmarks / Es & Os | Question | Answer | Notes |
|---|---|---|---|
| | 4b | No, it is not possible (pupil should provide a satisfactory explanation) 5 sweets to each bag 2 sweets left over | **On track**<br>• Understands that 'share equally' means give the same amount to each.<br>• Understands that there will be a remainder in the answer and can explain why.<br>• Can represent and solve the problem using a method of their choice, e.g. concrete materials, drawings or diagrams, or using known facts, i.e. $(5 \times 5) + 2 = 27$ or $27 \div 5 = 5$ r 2.<br><br>**Further learning required**<br>• May be unable to represent and solve the problem using concrete materials and/or drawings.<br>• May be able to solve the problem but have difficulty writing a number sentence to match how they solved it. |
| | 4c | 9 r 1<br>9 r 1 | **On track**<br>• Understands the division symbol.<br>• Uses knowledge of multiplication facts or repeated addition or subtraction to find the solution.<br><br>**Further learning required**<br>• May not understand the division symbol.<br>• May be unsure of multiplication and division facts and rely on concrete materials or drawings to solve the problems. |
| | 4d | 24 finalists<br>Pupils should show how they worked the answer out | **On track**<br>• Identifies multiplication as the required operation.<br>• Can represent and solve the problem using a method of their choice, e.g. concrete materials, drawings or diagrams, skip counting in 8s (repeated addition) or known facts.<br><br>**Further learning required**<br>• May be unable to represent and solve the problem or does so incorrectly.<br>• The above methods show a progression in the way a child thinks about a multiplication problem. Pupils dependent on counting in ones, using concrete materials or drawings, need practice in developing their ability to recall multiplication facts through partitioning arrays and skip counting. |

| Topic – benchmarks / Es & Os | Question | Answer | Notes |
|---|---|---|---|
| | 4e | 42<br>Array shows 7 rows of 6 (accept 6 rows of 7) | **On track**<br>• Understand the connection between an array and multiplication.<br>• Knows that they can partition the array to make calculation easier, e.g. '7 times 6 is the same as 6 sixes plus 6 more'.<br>**Further learning required**<br>• May not know how to draw an array.<br>• May confuse the rows and columns in an array.<br>• May miscount if using skip counting.<br>• May not understand that multiplication is commutative. |
| | 4f | 900 seeds | **On track**<br>• Uses knowledge of place value and repeated addition or counting in hundreds to solve the problem.<br>**Further learning required**<br>• May not understand that numbers get 100 times bigger when multiplied by 100.<br>• May talk about 'adding zeros' when multiplying by 100. |
| | 4g | 3<br>5 | **On track**<br>• Uses known facts and place value understanding to calculate accurately.<br>**Further learning required**<br>• May not understand that numbers get 100 times smaller when divided by 100.<br>• May talk about 'removing zeros' when dividing by 100. |
| | 4h | $7 \times 5 = 35$ or $5 \times 7 = 35$<br>$9 \times 2 = 18$ or $2 \times 9 = 18$ | **On track**<br>• Uses the inverse relationship between multiplication and division to write the corresponding multiplication number sentences.<br>**Further learning required**<br>• May not understand the inverse relationship between multiplication and division. |
| | 4i | $7 \times 2 = 14$<br>$2 \times 7 = 14$<br>$14 \div 2 = 7$<br>$14 \div 7 = 2$ | **On track**<br>• Understands the inverse relationship between multiplication and division and can use this knowledge to generate a family of four related facts.<br>**Further learning required**<br>• May not understand the inverse relationship between multiplication and division.<br>• May be able to write two multiplication facts but find generating division facts more challenging. |

| Topic – benchmarks / Es & Os | Question | Answer | Notes |
|---|---|---|---|
| | 4j | 5<br>Satisfactory explanation given, e.g. 'Because 9 × **5** = 45; **5** × 9 = 45; 45 ÷ 9 = **5**' | **On track**<br>• Uses known multiplication and division facts to find the unknown and can justify their answer.<br>**Further learning required**<br>• May be unable to write a number sentence to match their thinking. |
| | 4k | 300<br>Pupils should show how they worked it out | **On track**<br>• Uses place value understanding and known multiplication/division facts to calculate accurately.<br>• May link division and fractions and so find half of 600.<br><br>**Further learning required**<br>• May be unable to choose a strategy to solve the problem or have difficulty providing a satisfactory explanation, e.g. talks about 'removing zeros'. |
| | 4l | 10 minibuses | • May skip count in 20s on a number line or draw a bar model to help work out the answer.<br>• May use a two-step process to simplify the calculation, e.g. 200 ÷ 2 ÷ 10.<br><br>**Further learning required**<br>• May not understand the word problem or interprets it incorrectly, e.g. may multiply rather than dividing.<br>• May be unable to divide by a double digit or lack a two-step strategy. |
| | 4m | 86 + (5 × 9)<br>86 + 45 = 131<br>240 − 131 = 109<br><br>£109 | **On track**<br>• Correctly interprets the problem and chooses the correct operations to solve it.<br>• Each step in the problem-solving process is clearly shown.<br>**Further learning required**<br>• May struggle to understand and represent the word problem.<br>• May choose the wrong operation(s) and/or make calculation errors.<br>• May be unable to write a number sentence to show how they worked out each step. |

| Topic – benchmarks / Es & Os | Question | Answer | Notes |
|---|---|---|---|
| **Fractions, decimal fractions and percentages** MNU 1-07a MNU 1-07b MTH 1-07c | 5a | £9 Child draws a bar with a value of £27 divided into three equal parts. 11 glue sticks Child draws a bar with a value of 44 divided into four equal parts. | **On track** <br> • Connects finding a fraction of a quantity with division. <br> • Understands how to represent a problem as a bar model and uses known facts or a strategy, such as skip counting, to solve it. <br> **Further learning required** <br> • May not understand how to draw or use a bar model. <br> • May not understand how to find unit fractions of quantities. <br> • May make calculation errors. <br> • May be unsure of multiplication and division facts and fall back on a counting strategy. |
| | 5b 5c | Finlay scored 9 goals 8 boys were at the party | **On track** <br> • Can interpret the word problems and choose an appropriate strategy (such as those listed above) to solve them. <br> **Further learning required** <br> • See Further learning required section for question 5a. |
| | 5d | $\dfrac{1}{2}, \dfrac{1}{4}, \dfrac{1}{5}, \dfrac{1}{6}, \dfrac{1}{8}, \dfrac{1}{10}$ | **On track** <br> • Understands that the more parts a whole is divided into, the smaller each part will be. <br> • Understands what the numerator and denominator represent in unit fractions. <br> • Can use knowledge of unit fractions to order simple fractions from largest to smallest. <br> **Further learning required** <br> • May think that the larger the denominator, the larger the fraction and so wrongly order the fractions from smallest to largest. |

| Topic – benchmarks / Es & Os | Question | Answer | Notes |
|---|---|---|---|
| | 5e | 16 sixteenths<br>8 sixteenths<br><br>5f<br>18 eighteenths<br>6 eighteenths | **On track**<br>• Can use pictorial representations to identify equivalent fractions.<br><br>**Further learning required**<br>• May not know what the term equivalent fractions means.<br>• May be unable to work out the value of the unknown parts as sixteenths and eighteenths. |
| | 5g<br><br>5h | $\dfrac{2}{4}$ or $\dfrac{1}{2}$<br><br>$\dfrac{3}{8}$ | **On track**<br>• Understands how to divide a whole into equal fractions and how to complete a whole given a fraction of it.<br>If the answer $\dfrac{1}{2}$ is given then the pupil understands that $\dfrac{2}{4}$ and $\dfrac{1}{2}$ are equivalent.<br><br>**Further learning required**<br>• May not equate 'quarters' with four equal parts and so be unable to identify how much pizza each child gets. |
| **Money**<br>MNU 1-09a<br>MNU 1-09b | 6a | £18 and 76p OR £18.76 (but NOT £18.76p)<br>£10 note, £5 note, £2 coin, £1 coin, 50p, 20p, 5p, 1p<br><br>£6 and 90p OR £6.90 (but NOT £6.90p)<br>£5 note, £1 coin, 50p, 20p, 20p<br><br>£15 and 8p OR £15.08 (but NOT £15.08p)<br>£10 note, £5 note, 5p, 2p, 1p | **On track**<br>• Can read amounts of money written in pence and records each amount in pounds and pence.<br>• Correctly identifies the least number of notes and coins that can be used to make each amount.<br><br>**Further learning required**<br>• May interpret zero incorrectly, e.g. writes £6 and 9p and/or £15 and 80p OR<br>• May include the zero in the last example, e.g. writes £15 and 08p.<br>• May make each amount with notes and coins but not use the least number of coins. |
| | 6b<br><br>6c | £2.90 × 3 = £8.70<br>Yes, Tony has enough money<br><br>No, Olivia's mum needs another 10p | **On track**<br>• Uses a two-step process, i.e. find the totals by multiplying and/or adding then compares their answers with each bank total to decide if each character has enough money.<br><br>**Further learning required**<br>• May confuse pounds and pence at some point in the calculation.<br>• May not complete the question to state if there is sufficient money. |

| Topic – benchmarks / Es & Os | Question | Answer | Notes |
|---|---|---|---|
| | 6d | Finlay £5.67<br>Nuria £10 or £10.00<br>Isla £3.31<br><br>Isla and Finlay have £8.98 in total<br><br>Overall total £18.98<br>£10 note, £5 note, £2 coin, £1 coin, 50p, 20p, 20p, 5p, 2p, 1p | **On track**<br>• Records each amount of money correctly using decimal notation.<br>• Chooses an appropriate strategy to calculate each total.<br>• Correctly draws the overall total using the least number of notes and coins (allow basic drawing and follow on from their previous answers).<br>**Further learning required**<br>• May use decimal notation incorrectly.<br>• May confuse pounds and pence at some point in the calculation.<br>• May not complete the question as the pupil is unused to longer 2- or 3-step problems. |
| | 6e | £10 – £6.85 = £3.15 change | **On track**<br>• Calculates the correct change using their preferred method, e.g. by counting on or back on a number line or using bonds to 100.<br>**Further learning required**<br>• May start at the wrong place on the number line or miscount when totaling the jumps.<br>• May use bonds to 100 incorrectly, e.g. thinks 80 + 20 = 100; 5 + 5 = 10 and arrives at an incorrect answer of £3.25.<br>• May use an inappropriate strategy such as a standard algorithm. |
| **Time**<br>MNU 1-10a<br>MNU 1-10b<br>MNU 1-10c | 7a | i  Sunday<br>ii  13th<br>iii  14th February (accept Valentine's Day) | **On track**<br>• Can read information from a standard calendar.<br>• Knows how to write the date (allow for simple spelling errors).<br>**Further learning required**<br>• May not understand how to read a calendar.<br>• Struggles with the concept of 'the second Wednesday/Thursday'. |
| | 7b | i  Stopwatch<br>ii  Alarm clock<br>iii  Calendar<br>iv  Watch / clock<br>    Accept suitable alternatives, e.g. phone | **On track**<br>• Identifies a suitable device for measuring time in each example. |

| Topic – benchmarks / Es & Os | Question | Answer | Notes |
|---|---|---|---|
| | 7c | Correct hands on analogue clocks showing:<br>6:00<br>12:30<br>5:45<br>8:15 | **On track**<br>• Can draw the hands on an analogue clock to show each time.<br>**Further learning required**<br>• May be unable to use an analogue clock to tell the time.<br>• May confuse the short and long hands.<br>• May always show the hour hand pointing straight at the hour rather than in between the numbers for half-past times. |
| | 7d | 9:10<br>5:25<br>6:40<br>4:55 | **On track**<br>• Can convert between analogue and digital time correctly.<br>**Further learning required**<br>• May not understand the phrase 'digital form'.<br>• May reverse the position of hours and minutes, e.g. writes 10:9 for ten past nine. |
| | 7e | a.m.<br>Both<br>p.m.<br>p.m. | **On track**<br>• Uses a.m. and p.m. correctly.<br>**Further learning required**<br>• May not know the difference between a.m. and p.m. |
| | 7f | Weather – 19:30 / half past 7<br><br>Peeta Planet – 30 mins / half an hour<br><br>Basil Bay – 30 mins / half an hour<br><br>Bigville – end of News – 2 and a half hours / 2 hours 30 minutes | **On track**<br>• Can read a simple timetable.<br>• Accurately calculates durations in hours and half hours, mentally, using an empty number line or by making informal jottings.<br>**Further learning required**<br>• May misread the timetable.<br>• May attempt to calculate durations using an inappropriate method; for example, a standard algorithm. |

| Topic – benchmarks / Es & Os | Question | Answer | Notes |
|---|---|---|---|
| **Measurement**<br>MNU 1-11a<br>MNU 1-11b | 8a | Child draws two lines measuring:<br>5 cm<br>9.5 cm | **On track**<br>• Uses a ruler accurately to measure and record length in centimetres. Allow a tolerance of ±2 mm in question 8a.<br>• Can measure and record length in millimetres.<br>**Further learning required**<br>• The pupil may not start at zero on the ruler.<br>• The measurement may not be accurate.<br>• The pupil may confuse centimetres and millimetres . |
| | 8b | Seven and a half centimetres / $7\frac{1}{2}$ cm | |
| | 8c | 57mm | |
| | 8d | $1\frac{1}{2}$ kg (accept 1kg 500g) | **On track**<br>• Reads and records measurements on a scale.<br>**Further learning required**<br>• May not interpret the divisions on the scale correctly.<br>• May record the answer using incorrect notation. |
| | 8e | 6 kg | |
| | 8f | 3 litres and 3000 ml | **On track**<br>• Reads and records measurements on a scale.<br>**Further learning required**<br>• May not interpret the divisions on the scale correctly.<br>• May record the answer using incorrect notation. |
| | 8g | 45 squares | **On track**<br>• Understands the word 'area'.<br>• Can measure and record area by counting squares and half squares.<br>**Further learning required**<br>• May be unable to keep track of the squares counted.<br>• May be unable to combine half squares to make whole squares.<br>• May count half squares as whole squares. |
| | 8h | Draws two shapes on squared paper each with an area of 12 squares<br><br>Draws a shape with an area of $16\frac{1}{2}$ squares on squared paper | **On track**<br>• Creates shapes with a specified area.<br>**Further learning required**<br>• May be unable to create shapes of a specified area.<br>• May not keep track of the squares counted.<br>• May create a shape that is complex and difficult to track the area of. |

| Topic – benchmarks / Es & Os | Question | Answer | Notes |
|---|---|---|---|
| | 8i | $5 \times 3 = 15$ or $3 \times 5 = 15$ | **On track**<br>• Interprets the information provided as '3 rows of 5' or '5 columns of 3' and can connect this with $3 \times 5$ or $5 \times 3$.<br>**Further learning required**<br>• May not understand how an array and area are connected to multiplication. |
| **Mathematics, its impact on the world, past, present, future**<br>MTH 1-12a | 9 | 二十八<br><br>十六 | **On track**<br>• Recognises different ways in which numbers are represented across the world.<br>• Can recognise place value, tens and ones, in an alternative number system that uses symbols.<br>**Further learning required**<br>• May not recognise the 10s and ones represented by the Chinese symbols. |
| **Patterns and relationships**<br>MTH 1-13b | 10a<br><br><br>10b | 3, 6, **9, 12**, 15, **18**<br>18, 12, 6, **0**<br><br>20, 24, 28<br>90, 105, 120 | **On track**<br>• Correctly works out the rule and completes the number patterns.<br>**Further learning required**<br>• May not be able to work out the rule used to create the sequence.<br>• May miscalculate. |
| | 10c | 10, 16, 22, 28, 34<br>180, 160,140,120,100<br>27, 39, 51, 63, 75 | **On track**<br>• Correctly works out the rule and continues the number patterns.<br>**Further learning required**<br>• May misinterpret the rules.<br>• May not understand the word 'term'.<br>• May make calculation errors, particularly when bridging a 10 or 100, e.g. 16 add 6; 28 add 6; 200 – 20; 27 +12, etc. |
| | 10d | 40, 35<br>Rule: subtract 5<br><br>35<br>Rule: add 3 | **On track**<br>• Correctly works out the rule used to complete the number pattern.<br>**Further learning required**<br>• May not know to look for the difference between the numbers. |

| Topic – benchmarks / Es & Os | Question | Answer | Notes |
|---|---|---|---|
| **Expressions and equations**<br>MTH 1-15a<br>MTH 1-15b | 11a<br>11b<br>11c<br>11d | 7<br>24<br>20<br>5 | **On track**<br>• Understands that the equals sign symbolises balance in equations.<br>• Uses known facts and/or the inverse relationship to solve the equations.<br><br>**Further learning required**<br>• May be unfamiliar with the format of the equations.<br>• May add 7 + 14 in question a and subtract 9 from 11 in question c. |
| | 11e | Circle equals 5<br>Triangle equals 4<br>Square equals 9 | **On track**<br>• Uses algebraic thinking to understand that the shapes can represent values.<br>• Uses doubling/halving, known facts or other strategies to solve for the missing values.<br><br>**Further learning required**<br>• May not understand that the shapes represent values.<br>• May have difficulty reasoning about the problem.<br>• May rely on a less efficient strategy such as counting in ones. |
| **2D shapes and 3D objects**<br>MTH 1-16a | 12a | Cuboid<br>6 faces<br>8 vertices<br><br>Triangular prism<br>5 faces<br>6 vertices<br><br>Cylinder<br>3 faces<br>0 vertices<br><br>*Pupils may use some or all of the following words when describing faces: rectangle, triangle, circle, flat, curved | **On track**<br>• Identifies 3D objects correctly and describes their properties, using the words faces and vertices.<br><br>**Further learning required**<br>• May not understand what the face of a shape is.<br>• May be unfamiliar with the word vertices.<br>• May not know the names of the 3D objects shown. |
| | 12b | Cone | **On track**<br>• Identifies 3D objects from a plan and side view.<br>**Further learning required**<br>• May only be able to identify 3D shapes when presented with real objects. |

| Topic – benchmarks / Es & Os | Question | Answer | Notes |
|---|---|---|---|
| **Angles, symmetry and transformation**<br>MTH 1-17a<br>MTH 1-18a<br>MTH 1-19a | 13a<br>13b | 90 degrees<br>B, D | **On track**<br>• Knows the properties and size of a right angle.<br>• Can identify right angles in 2D drawings.<br>**Further learning required**<br>• May only recognise right angles when they are in a standard orientation. |
| | 13c | F2 | **On track**<br>• Can follow directions using words associated with angles and turning, e.g. right 90, left 90, forwards, backwards.<br>**Further learning required**<br>• May confuse left and right.<br>• May not understand that right and left indicate right angle turns.<br>• May miscount squares. |
| | 13d | i  Finlay's house to Nuria's house:<br>3 squares east,<br>1 square south,<br>1 square east,<br>2 squares south,<br>2 squares east,<br>1 square south<br><br>ii  Amman's house to Finlay's house:<br>1 square east<br>3 squares north<br><br>Other routes are possible | **On track**<br>• Knows and can use the compass points north, south, east and west.<br>**Further learning required**<br>• May confuse compass points, in particular east and west. |
| | 13e | Many answers are possible | **On track**<br>• Correctly creates a symmetrical design with one line of symmetry using two colours.<br>**Further learning required**<br>• May copy the design they have created but not reflect it around the line of symmetry. |
| | 13f | Pupil draws an object of their own choice with more than one line of symmetry | **On track**<br>• Understands some shapes have more than one line of symmetry.<br>• Correctly creates a symmetrical design with more than one line of symmetry.<br>**Further learning required**<br>• May have problems identifying where the lines of symmetry are. |

| Topic – benchmarks / Es & Os | Question | Answer | Notes |
|---|---|---|---|
| **Data handling and analysis**<br>MNU 1-20a | 14a | <table><tr><td>Month</td><td>Tally/Total</td></tr><tr><td>January</td><td>1</td></tr><tr><td>February</td><td>2</td></tr><tr><td>April</td><td>1</td></tr><tr><td>May</td><td>3</td></tr><tr><td>August</td><td>1</td></tr><tr><td>September</td><td>1</td></tr><tr><td>December</td><td>1</td></tr></table><br>Some pupils may not use a table or may not put months in calendar order, but accept an organised list with tallies or totals<br><br>Accept a statement such as, 'There are more birthdays in May than any other month' or 'May and February have more than one birthday' or a similar correct statement based on the data | **On track**<br>• Can organise data into a list or table and can find tallies and totals.<br>• Can write a true statement based on the data.<br>**Further learning required**<br>• Does not know how to organise data in a list or table.<br>• May not find totals and instead compare an estimate of the data instead of the actual quantities.<br>• Cannot write a true statement about the data presented and instead gives an answer based on their own personal experiences, e.g. 'January is the best month for a birthday because that is when my birthday is'. |

| Topic – benchmarks / Es & Os | Question | Answer | Notes |
|---|---|---|---|
| | 14b | | |

| | Odd numbers | Even numbers |
|---|---|---|
| Less than ten | 1, 3, 5, 7, 9 | 2, 4, 6, 8 |
| Ten or more | 11, 13, 15, 17, 19 | 10, 12, 14, 16, 18, 20 |

**On track**
- Uses data accurately to create a Carroll diagram.
- Understands the terms, odd, even, more than and less than.

**Further learning required**
- May not understand what a Carrol diagram is.
- May not understand the mathematical terms.

**14c**

Multiples of 5

5, 15, 25, 35, 45, 55

Multiples of both 5 and 10

10, 20, 30, 40, 50, 60, 70, 80, 90, 100

Multiples of 10 only

**On track**
- Uses data accurately to create a Venn diagram.
- Understands the term multiples.

**Further learning required**
- May not understand what a Venn diagram is or how to complete shared values.
- May not understand the mathematical terms.

**14d**

True
False
False
True

**On track**
- Interprets the data in the graph correctly to answer questions.

**Further learning required**
- May not understand the language of true and false.
- May struggle to read the values from the graph.

| Topic – benchmarks / Es & Os | Question | Answer | Notes |
|---|---|---|---|
| | 14e | 100<br>Thursday<br>30 more | **On track**<br>• Interprets the data in the graph correctly to answer questions.<br>**Further learning required**<br>• May not understand the language of difference or how many more.<br>• May struggle to read the values from the graph. |
| | 14f | <br>Pupils may correctly draw a bar graph with the bars horizontal. | **On track**<br>• Can create a bar graph from the information provided:<br>• the graph has a suitable title<br>• both axes are appropriately labelled<br>• the values are correct for each bar<br>• bars are equally spaced<br>**Further learning required**<br>• May draw a pictogram or similar replacement.<br>• May not include all labels.<br>• The bars may not be accurately drawn.<br>• Bars may be unequal in width and/or joined together. |
| **Ideas of chance and uncertainty**<br>MNU 1-22a | 15a | Certain<br>Possible<br>Impossible | **On track**<br>• Understands the vocabulary of probability and can use it to make judgements about the likelihood of events occurring. |
| | 15b | Will happen<br>Could happen<br>Will not happen<br>Could happen<br><br>*Alternative answers are acceptable, providing the pupil can provide an explanation that proves they understand the meaning of the vocabulary in bold. | **Further learning required**<br>• May not understand the mathematical interpretation of the chance words, e.g. refusing to use 'impossible' in the belief that 'nothing is impossible'. |

# End of first level assessment 1

## The journey

1. Amman and his family are going to visit Gran. Amman loves visiting Gran because she tells him all about the way she uses maths and numbers in her job. What job might Gran have?
Explain the reason for your answer.

2. 'We can't take lots of bags,' Dad says.
'That's okay. My bag is 7 kg and 500 g' says Amman.
'Mine is only 500 g,' says Mum.
'If we add the mass of all our bags, we would have 20 kg,' says Dad.
How heavy is Dad's bag?

3. The mile counter on Dad's car shows how far the car has travelled in the last month. It shows this number:

Gran's house is about 89 miles away. Amman estimates that the counter will say 980 when they arrive.

Is his estimate reasonable? Explain your thinking.

4. a When they arrive in Gran's street, Amman looks for Gran's house, which is 236 Park Road. He notices that he passes houses 212, 214, 216.

Continue the number sequence until Gran's house and write down the rule.

b   Amman thinks it would be more interesting if the rule was add 3.

What would the house numbers be for the 5 houses before Gran's house if that was the rule?

5. a   On the way home, Amman watches the mile counter again.

He decides to eat a sweetie every 10 miles, starting when they have travelled 10 miles. The journey home is exactly 90 miles. What number will show on the counter each time Amman eats a sweet?

b   How many sweets does Amman eat on the journey home? How did you work it out?

# End of first level assessment 2

**You will need:**
- Blank watch / clock face
- Carroll diagram
- Plan of the play park
- Angles table
- Coins table

## The park

1. The children are at the park. Mum shows Nuria the time on her phone. It says:

15:00

Mum tells Nuria she can stay at the park for an hour.

a Show the time Nuria must leave on a clock or watch like this one.

b Is Nuria in the park in the morning, the afternoon or the evening? Justify your answer.

2. The park has these things in it:

   a roundabout      a slide      a seesaw      a climbing frame.

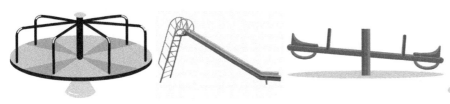

   Complete the Carroll diagram like this one by writing or drawing.

|  | **Is round** | **Is not round** |
|---|---|---|
| Higher than 2 metres |  |  |
| Not higher than 2 metres |  |  |

3. a   The play park is shaped like this:

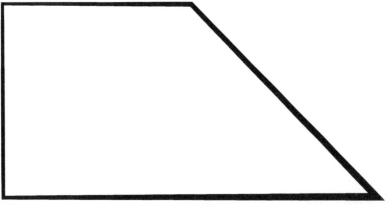

   Mark the right angles on the plan like the one above.

   b   The park has a large football pitch. It has a flower bed and a pond.

   football pitch      flower bed     pond

 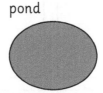

   Complete the table like the one below. Put a tick in the correct place for each object.

|  | **All right angles** | **Some right angles** | **No right angles** |
|---|---|---|---|
| football pitch |  |  |  |
| flower bed |  |  |  |
| pond |  |  |  |

4.  a  The ice cream van comes to the park. An ice cream cone costs 76p.

Amman says he can pay with six coins, a 20p, a 20p, a 10p, a 10p, a 15p and a 1p.

Nuria shakes her head and says, 'That's not possible.'

Who is correct? Explain your thinking.

b  All four children buy an ice cream and give the exact amount. However, each child pays for it in a different way.

Show the coins that each child uses on a table like the one below.

| Finlay | |
| --- | --- |
| Isla | |
| Nuria | |
| Amman | |

# End of first level assessment 2

5. a    Some children come to play on the football pitch. There are 17 children and they get into two teams.

      Is this fair? Explain your thinking.

   b    One of the players has to go home because he has hurt his foot. The other football players stop for a break and each player is given one quarter of an orange. Draw the quarters and work out how many whole oranges he started with.

# End of first level assessment 3

**You will need:**
- Blank paper
- Shapes for tiling

## The garden

1. Mum hides a treat for Isla and Nuria. She tells them to start at the garden gate and go forward 3 steps, turn right 90°, forward 2 steps, turn right 90°, forward 2 steps, turn left 90° and then forward 1 step.

   a. Where is the treat hidden?

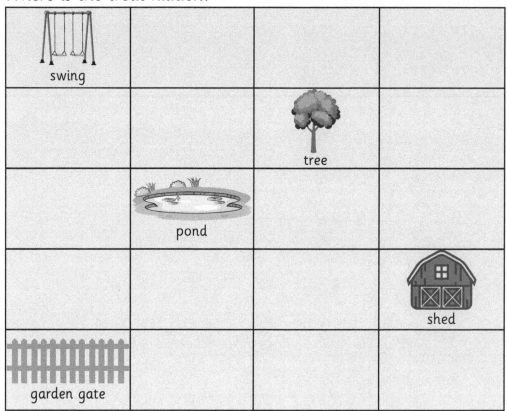

   Isla and Nuria hide a treat for Mum by the tree. They both give Mum directions using north, south, east, west but their directions are different. Mum starts at the garden gate and goes one step north ...

   b  What does Nuria write? What does Isla write?

# End of first level assessment 3

2. The path in the garden is straight and is made from tiles like these  and ◣ .

Use the paper, square and triangle your teacher gives you to show how the path could be tiled.

3. The girls help Mum to plant some seeds. There are 20 seeds in each packet.

SEEDS    SEEDS    SEEDS

How many seeds do they have?

4. a   Mum said, 'Plant the seeds in 8 rows with 10 seeds in each.'

Is this possible? Explain your thinking.

b   Show how the girls could plant the seeds in rows of 5.

5. Gran comes into the garden. She has some plant pots and 23 plants. 'I want the pots to have the same number of flowers in each. If you need more plants, come and get me.'

The girls needed 2 more plants to make the pots the same. How many pots were there?

# End of first level assessment 4

## The school fair

It is the day of the School Fair and all the children are going.

1. Amman sees a stall with good prizes. The man on the stall tells him he can win a prize in two ways. He can roll a dice and get a six or he can toss a coin and get a head.

   Which way will be more likely to win a prize? With the dice or the coin? Justify your answer.

2. Isla plays a game where she has to throw a coin in a bucket.

   Is it most likely that the bucket is 2 km, 2 m or 2 mm from her? Explain your thinking

# End of first level assessment 4

3. Amman wins a prize at a sweetie game. He can take $\frac{1}{2}$ of the sweets or $\frac{1}{5}$ of the sweets.

Which fraction should he take to have the most sweets? Explain your thinking.

4. Nuria was asked to record the number of people that came to the fair. She recorded the number of people like this:

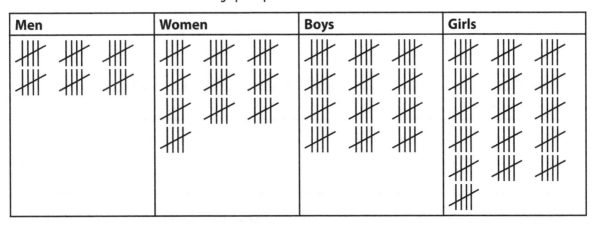

Draw a bar chart on squared paper to show this data.

# End of first level assessment 4

5. In the hall, the children can go on the stage and perform. Nuria does a magic trick.

   She asks her teacher to think of a two-digit number. The teacher thinks of 54.

   Nuria tells the teacher to double the number.

   She then tells the teacher to add 10 to the new number.

   Next the teacher has to half that new number.

   The teacher then has to take away the number she first thought of.

   Nuria tells her the answer is 5.

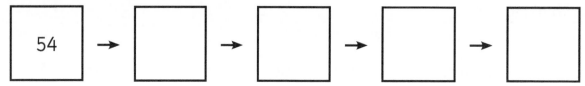

   a  Is Nuria correct?

   b  Choose a two-digit number and follow Nuria's instructions. What is the answer?

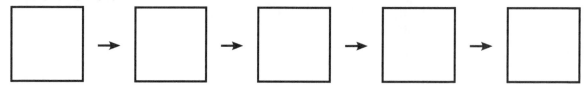

6. Finlay plays a game. There are three prizes. Inside each box there are some toys.

   **WHICH PRIZE HAS THE MOST?**

   C  <  A      C  >  B

   Which box should Finlay choose to have the most toys?

7. Finlay goes to the Pairs Game. He must find three pairs that each total 100. Finlay says this is impossible.

   Is he correct? Explain your answer.

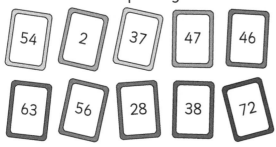

8. The school makes a display to show how much money was made at the fair. The display has three cards with a digit on each: 0, 4 and 7.

   **WE MADE**

   How much might the school have made? Show four different answers.

**The party**

1. Isla's birthday is on the 29th June and Finlay has been invited to her party. He is excited because his birthday is only 10 days later.

   What date is Finlay's birthday? Write it using numbers and words.

2. Finlay's mum has given him £10 to buy Isla a gift. She gave him seven coins.

   Show two different ways Mum might have done this.

3. Finlay picked two gifts from the items shown below.

   Slime £5.50    Fidget toy £1.49    Lipgloss set £8.99

   a  Which two things did Finlay buy with the £10 Mum gave him.

   b  How much money does Finlay have left?

4. a  Finlay puts his gifts in a box before wrapping it up. The box is a cuboid. The length of the box is 30 cm. The depth of the box is 5 cm. The width is half the length. What is the width of the box in centimetres?

   b  Finlay ties some ribbon all the way around the box as shown in the picture. He has 1 m 10 cm of ribbon.

# End of first level assessment 5

Does he have enough ribbon to put around the box and tie a bow?
Explain your thinking.

5. Finlay is going to the party with Amman. He needs 20 minutes to walk to Amman's house, where Amman is waiting, and then they need 15 minutes to walk to Isla's house. The party starts at 2:00 p.m.

What time does Finlay need to leave his house to get there on time?
Show how you worked out
your answer.

6. What is the grid reference of Isla's house?

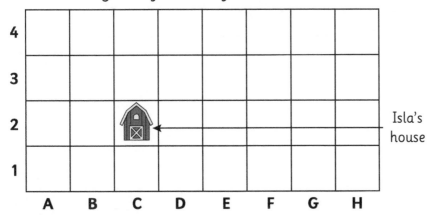

7. Isla's mum made a cake for the party. She opened a 1 kg bag of sugar and used half of it.

How many grams of sugar were left in the bag?

# End of first level assessment 5

8.  a  Isla's mum made milkshakes for the party. She measured out some milk to fill a 250 ml jug and had 750 ml left in the bottle.

    How many litres of milk were in the bottle when she opened it? Justify your answer.

    b  How many more jugs of milk can Isla's mum pour?

9.  The cake had 5 faces, 6 vertices and 9 edges.

    a  What shape was the cake?

    b  What shapes were the cake's 5 faces?

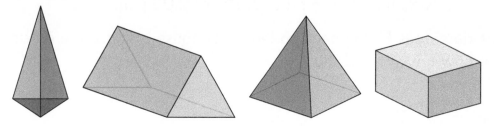

10. a  Mum also made cookies that were these shapes. Which shape has the least lines of symmetry?

    A          B          C

    b  Draw a cookie shape that has exactly 2 lines of symmetry.

# End of first level assessment 6

**You will need:**
- Venn diagram

## The school

1. a It is Monday. When the children go into the classroom, their teacher has a problem for them. She gives each child a card and asks them to get into order from smallest to largest.

What order will the fractions go in?

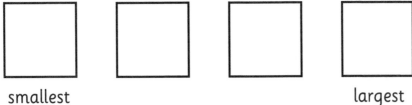

smallest                                           largest

b This time, the teacher gives the children different cards. They arrange themselves in this order.

Are they in order from smallest to largest? Explain your answer.

2. The teacher puts a balance pan on the table. It shows this:

How heavy is this bag? 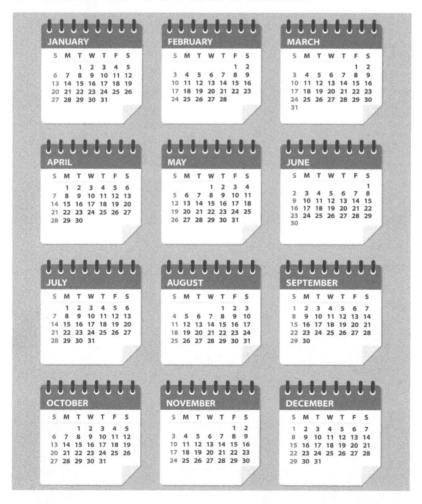 Show how you worked it out.

3. The teacher shows the class a calendar.

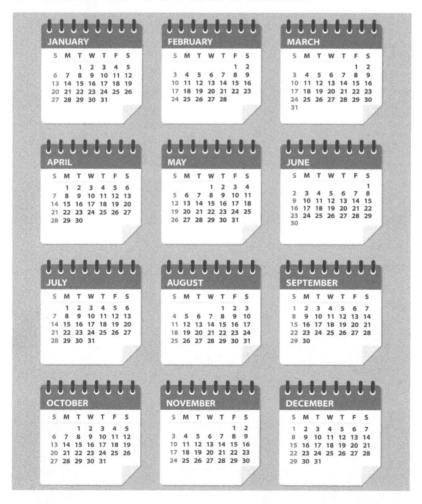

She tells the children they are going on a day trip on the first Monday in October.

a What date are the children going on a trip?

b What season will it be?

4. The children want to know where they are going on the trip. 'It might be the beach, it might be the woods or it might be France!' says the teacher. Which is least likely and why?

5. The Venn diagram shows what the children do at lunch time.

a What does Nuria do at lunch time?

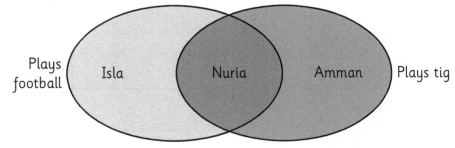

Plays football    Isla    Nuria    Amman    Plays tig

b Finlay does not play tig or play football at lunch time. Draw Finlay on the Venn diagram like this one.

c Ailsa plays football but not tig. Draw Ailsa on the Venn diagram like this one.

# End of first level assessments: answers and marking guidance

## Assessment 1: The journey

| Assessment | Topic | Question |
|---|---|---|
| Q1 | Mathematics – its impact on the world, past, present and future<br><br>MTH 1-12a | **On track**<br>Child can identify any job that uses numbers and give a reasonable explanation. If unsure, ask child to explain their thinking.<br><br>**Review**<br>Ensure child has exposure to the place of number in the lives of those around them and in their community. |
| Q2 | Measurement<br><br>MNU 1-11a | **On track**<br>12kg<br><br>**Review**<br>Identify whether difficulty lies in understanding of relationship between grams and kilograms or in calculation. Ensure child can extract information for multi-step questions and break them down logically. |
| Q3 | Estimating and rounding<br><br>MNU 1-01a | **On track**<br>No. Accept answers that show understanding that 89 is nearly 90 or that 89 is nearly 100 and 980 is almost 200 more than 784 and is therefore too high an estimation.<br><br>**Review**<br>Ensure child can round numbers to nearest 10 and 100 and can add 100 to 3-digit numbers. |
| Q4a | Patterns and relationships<br><br>MTH 1-13b | **On track**<br>218, 220, 222, 224, 226, 228, 230, 232, 234, 236<br>Rule: Add 2<br><br>**Review**<br>Ensure child can count orally in twos. Encourage child to articulate the rule when counting orally backwards and forwards. |
| Q4b | Patterns and relationships<br><br>MTH 1-13b | **On track**<br>221, 224, 227, 230, 233<br><br>**Review**<br>Ensure child can count orally in threes. Encourage child to articulate the rule when counting orally backwards and forwards. |
| Q5a | Number and number processes<br><br>MNU 1-02a | **On track**<br>883, 893, 903, 913, 923, 933, 943, 953, 963<br><br>**Review**<br>Ensure child can skip count in tens from any number. |

# End of first level assessments: answers and marking guidance

| Assessment | Topic | Question |
|---|---|---|
| Q5b | Number and number processes<br><br>MNU 1-02a | **On track**<br>9 sweets. Child may say 90 divided by 10 is 9, explain that they counted from 873 in tens and knew to stop at 963 because they had counted ten 9 times (perhaps 'double-counting' on their fingers), or may say they counted how many numbers they had written down, equating one number with 1 sweet.<br><br>**Review**<br>Encourage child to act out or visualise the problem. |

# End of first level assessments: answers and marking guidance

## Assessment 2: The park

| Assessment | Topic | Question |
|---|---|---|
| Q1a | Time<br><br>MNU 1-10a | **On track**<br>Correctly draws 4 o'clock with appropriate hand length.<br><br>**Review**<br>Check that child understands 24-hour clock and the relationship to 12 hour times. |
| Q1b | Time<br><br>MNU 1-10a | **On track**<br>Afternoon. Child explains that 15:00 is the same as 3 o'clock in the afternoon. May mention digital clock displays.<br><br>**Review**<br>Revisit 24-hour times. Check child's understanding of parts of the day. |
| Q2 | Data and analysis<br><br>MNU 1-20a | **On track**<br>Correctly completes Carroll diagram.<br><br><table><tr><td></td><td>**Is round**</td><td>**Is not round**</td></tr><tr><td>Higher than 2 metres</td><td></td><td>slide<br>climbing frame</td></tr><tr><td>Not higher than 2 metres</td><td>roundabout</td><td>seesaw</td></tr></table><br>**Review**<br>Give child experiences of estimating measurements in the environment if required. If child understands measurement but cannot complete Carroll diagrams, give further opportunities in this area. |
| Q3a | Angle, symmetry and transformation<br><br>MTH 1-17a | **On track**<br>Marks two right angles only: top left and bottom left of shape.<br><br>**Review**<br>Ensure child has experience of checking for right angles using a right-angle tester. |

# End of first level assessments: answers and marking guidance

| Assessment | Topic | Question |
|---|---|---|
| Q3b | Angle, symmetry and transformation<br><br>MTH 1-17a | **On track**<br>As table.<br><br>_table below_<br><br>**Review**: Give child opportunities to find right angles using a right-angle tester. |
| Q4a | Money<br><br>MNU 1-09a | **On track**<br>Nuria is correct because there is no 15p coin.<br><br>**Review**<br>Ensure child has experience of handling coins and knows the value of coins in circulation. |
| Q4b | Money<br><br>MNU 1-09b | **On track**<br>There are multiple ways to show 76p. Ensure that each way is accurate. Do not accept the same coins written or drawn in a different order.<br><br>**Review**<br>Ensure child can partition 2-digit numbers in different ways and create the same amount using different combinations of coins in circulation. |
| Q5a | Number and number processes<br><br>MNU 1-03a | **On track**<br>No, because one team would have an extra player.<br><br>**Review**<br>Ensure child understands sharing as a method of dividing. |
| Q5b | Fractions, decimal fractions and percentages<br><br>MNU 1-07a | **On track**<br>Child draws 16 quarters and identifies 4 whole oranges.<br><br>**Review**<br>Check child understands that quarters are a fourth of the whole. Investigate how many fractional parts are the same as a whole. |

Table for Q3b:

|  | All right angles | Some right angles | No right angles |
|---|---|---|---|
| football pitch | ✓ |  |  |
| flower bed |  |  | ✓ |
| pond |  |  | ✓ |

# End of first level assessments: answers and marking guidance

## Assessment 3: The garden

| Assessment | Topic | Question |
|---|---|---|
| Q1a | Angle, symmetry and transformation<br><br>MTH 1-17a | **On track**<br>The shed.<br><br>**Review**<br>Ensure child can tell right from left and understands how right and left depend upon the direction that a person is facing. Ensure child understands 90° as a quarter turn. |
| Q1b | Angle, symmetry and transformation<br><br>MTH 1-17a | **On track**<br>There are multiple possible answers. For example, 1 step north, 2 east, 2 north or 1 step north, 2 steps north, 2 east.<br><br>**Review**<br>Ensure child knows compass point directions and understands that they remain static regardless of the direction a person faces. |
| Q2 | Properties of 2D shapes and 3D objects.<br><br>MTH 1-16b | **On track**<br>Child creates pattern with no gaps using both shapes.<br><br>**Review**<br>If pattern has gaps, ensure the child has experience of handling representations of 2D shapes and explore rotating them and viewing them from different perspectives. |
| Q3 | Number and number processes<br><br>MNU 1-03a | **On track**<br>60<br><br>**Review**<br>Ensure child appreciates that the problem requires multiplication/repeated addition and has a suitable strategy for finding the answer, e.g. skip counting or using known facts and place value. |
| Q4a | Number and number processes<br><br>MNU 1-03a | **On track**<br>No. The children need 80 seeds / 4 packets and they only have 60 seeds / 3 packets. Check answer against Question 3 to allow for incorrect answer at Q3 being correctly interpreted in Q4.<br><br>**Review**<br>Ensure child can skip count in tens. |

# End of first level assessments: answers and marking guidance

| Assessment | Topic | Question |
|---|---|---|
| Q4b | Number and number processes<br><br>MNU 1-03a | **On track**<br>Child draws 12 rows of 5 or 5 rows of 12 or child writes $5 \times 12 = 60$.<br><br>**Review**<br>Explore sharing and grouping of objects. Draw attention to the relationship between division and multiplication. |
| Q5 | Number and number processes<br><br>MNU 1-03a | **On track**<br>5 pot plants. Do not accept 25 pots as the question makes clear there must be more than 1 plant in each pot.<br><br>**Review**<br>Ensure child can visualise / represent the problem and can reason that, if the girls had 23 plants and got 2 more then they would have 25. Check the child has a strategy for working out the number of plants in each pot, for example by grouping concrete materials, drawing or using known facts. |

# End of first level assessments: answers and marking guidance

## Assessment 4: The school fair

| Assessment | Topic | Question |
|---|---|---|
| Q1 | Ideas of chance and uncertainty<br><br>MNU 1-22a | **On track**<br>Identifies tossing a coin and correctly identifies that there are only two possible outcomes whereas there are six possible outcomes when rolling a dice.<br><br>**Review**<br>Carry out practical investigations to identify possible outcomes. |
| Q2 | Measurement<br><br>MNU 1-11a | **On track**<br>Child demonstrates an understanding of the impossibility of throwing a distance of 2 km and the closeness of 2 mm.<br><br>**Review**<br>Explore measurement in a variety of practical settings. Make comparisons visible. |
| Q3 | Fractions, decimal fractions and percentages<br><br>MNU 1-07b | **On track**<br>Child explains that half of 20 is 10 and a fifth of 20 is 4 so $\frac{1}{2}$ of the quantity is more.<br><br>**Review**<br>Ensure child understands that the denominator denotes the number of equal shares and not the amount in each share. Investigate finding different fractions of the same amount using concrete materials. |
| Q4 | Data and analysis<br><br>MNU 1-20b<br>MTH 1-21a | **On track**<br>Correctly counts the tally marks to arrive at the following totals: men 30, women 50, boys 60, girls 80. Plots graph appropriately labelling each axis and giving a title. Uses a suitable scale.<br><br>**Review**<br>Check child is familar with tally marks and can skip count in fives. Discuss a variety of bar charts and identify common factors, e.g. labelled axes, title. Allow opportunities to talk about graphs where one unit represents more than one data value. |
| Q5a | Number and number processes<br><br>MNU 1-03a | **On track**<br>Yes 54 – 108 – 118 – 59 – 54<br><br>**Review**<br>Identify whether doubling, addition, halving or subtraction is the point of error. Ensure children have further experience in area of weakness. |

# End of first level assessments: answers and marking guidance

| Assessment | Topic | Question |
|---|---|---|
| Q5b | Number and number processes<br><br>MNU 1-03a | **On track**<br>5. Ask children to compare start and finish numbers. Can they figure out why this happens?<br><br>**Review**<br>Identify whether doubling, addition or halving is the point of error. Ensure child has further experience in area of weakness. |
| Q6 | Expressions and equations<br><br>MTH 1-15a | **On track**<br>A. Check that the child can justify their answer and haven't simply taken a lucky guess.<br><br>**Review**<br>Ensure child understands symbols meaning greater than and less than. Child may need practice in making jottings as they extract information. |
| Q7 | Number and number processes<br><br>MNU 1-03a | **On track**<br>No because 37 + 63, 54 + 46 and 72 + 28 all make 100.<br><br>**Review**<br>Look at the strategies the child has used to calculate additions. Encourage partitioning strategies. |
| Q8 | Number and number processes<br><br>MNU 1-02a | **On track**<br>£407, £470, £740, £704<br><br>**Review**<br>If child records 047 and/or 074 discuss use of 0 in numbers and its purpose of being a place holder. |

# End of first level assessments: answers and marking guidance

## Assessment 5: The party

| Assessment | Topic | Question |
|---|---|---|
| Q1 | Time<br><br>MNU 1-10b | **On track**<br>Correctly identifies ten days from 29th June as 9th July and records it as ninth of July and 9th July.<br><br>**Review**<br>If the child is incorrect, check their knowledge of the sequence of months, the number of days in each month and whether the child understands that he/she should count on from 29, i.e. 29th should not be included in the count. |
| Q2 | Money<br><br>MNU 1-09b | **On track**<br>Shows two ways of making £10 with 7 coins, i.e. £2, £2, £2, £2, £1, 50p, 50p AND £2, £2, £2, £1, £1, £1, £1<br><br>**Review**<br>If the child has difficulty making £10 in different ways, they may need further exposure to using different coins to make the same amounts. Check their understanding of the relationship between pounds and pence. |
| Q3a | Money<br><br>MNU 1-09a | **On track**<br>Correctly identifies slime (£5.50) and fidget toy (£1.49) as the only two possibilities.<br><br>**Review**<br>Calculates either addition or subtraction part of the question incorrectly. Check understanding and method used to identify two items. |
| Q3b | Money<br><br>MNU 1-09a | **On track**<br>Adds £5.50 and £1.49, calculates change and records as £3.01.<br><br>**Review**<br>Records change as 301p or £3.01. Ensure child can convert pence to pounds and recognises this as the standard way of representing money. |
| Q4a | Fractions, decimal fractions and percentages<br><br>MNU 1-07b | **On track**<br>15 cm. The child reasons that half of 30 is 15.<br><br>**Review**<br>If incorrect, check understanding of language length and width. Check child understands the link between fractions and division and has a strategy for dividing whole numbers by 2. |

# End of first level assessments: answers and marking guidance

| Assessment | Topic | Question |
|---|---|---|
| Q4b | Number and number processes<br><br>MNU 1-03a | **On track**<br>No because the minimum amount of ribbon required (110 cm without bow) will be more than 110 cm as that is the sum of the sides which require ribbon $(2 \times 30$ cm$) + (2 \times 15$ cm$) + (4 \times 5$ cm$)$.<br><br>**Review**<br>If incorrect, check the child's ability to add multiples of 10, their knowledge of the fact that 100 cm = 1 m and their ability to work with depth, width and height in practical settings. |
| Q5 | Time<br><br>MNU 1-10a | **On track**<br>1:25 p.m. or twenty-five past one. Child gives a satisfactory explanation of how they worked it out.<br><br>**Review**<br>If incorrect, check the child's ability to tell the time and calculate simple durations of time by counting on or counting back. |
| Q6 | Angle, symmetry and transformation<br><br>MTH 1-18a | **On track**<br>C2<br><br>**Review**<br>Ensure child is reading horizontal before vertical. |
| Q7 | Measurement<br><br>MNU 1-11a | **On track**<br>500 g<br><br>**Review**<br>If incorrect, check understanding of the relationship between grams and kilograms. |
| Q8a | Measurement<br><br>MNU 1-11a | **On track**<br>1 litre. Child explains that 250 ml and 750 ml is 1000 ml, which is the same as 1 litre.<br><br>**Review**<br>Check that the child understands the relationship between litres and millilitres, i.e. that 1000 ml = 1 l. Check strategies available to the child for adding 3-digit multiples of 10. If necessary, give further experience of reading scales in millilitres and litres in a practical setting. |

# End of first level assessments: answers and marking guidance

| Assessment | Topic | Question |
|---|---|---|
| Q8b | Measurement<br><br>MNU 1-11a | **On track**<br>3<br><br>**Review**<br>If incorrect, check the child's ability to divide a 3-digit multiple of 10 by a single digit. If necessary, give practical measurement experiences of subtracting from a given amount. |
| Q9a | Properties of 2D shapes and 3D objects.<br><br>MTH 1-16a | **On track**<br>Second shape identified (triangular prism).<br><br>**Review**<br>Ensure child understands terms faces, vertices and edges. Have child handle shapes and explore them from a variety of perspectives. |
| Q9b | Properties of 2D shapes and 3D objects.<br><br>MTH 1-16a | **On track**<br>Triangles and rectangles. Child may draw three rectangles and two triangles.<br><br>**Review**<br>If child answers incorrectly, explore faces of common 3D shapes. Ensure child can name and identify 2D shapes. |
| Q10a | Angle, symmetry and transformation<br><br>MTH 1-19a | **On track**<br>Shape A (1 line of symmetry)<br><br>**Review**<br>Check understanding of what the term symmetry means. Ensure child has experience of folding shapes to demonstrate different lines of symmetry. |
| Q10b | Angle, symmetry and transformation<br><br>MTH 1-19a | **On track**<br>Accept any shape with only 2 lines of symmetry.<br><br>**Review**<br>Check understanding of what the term symmetry means. Ensure child has experience of folding shapes to demonstrate different lines of symmetry. |

# End of first level assessments: answers and marking guidance

## Assessment 6: The school

| Assessment | Topic | Question |
|---|---|---|
| Q1a | Fractions, decimal fractions and percentages<br><br>MNU 1-07a | **On track**<br>Orders $\dfrac{1}{10}$, $\dfrac{1}{5}$, $\dfrac{1}{4}$, $\dfrac{1}{2}$<br><br>**Review**<br>If child makes any error, give opportunities to explore different fractional parts of identical whole objects, practically. |
| Q1b | Fractions, decimal fractions and percentages<br><br>MNU 1-07a | **On track**<br>Yes. Child identifies that the children are in order and gives an answer showing that they understand that the more parts the same object is split into, the smaller each part will be.<br><br>**Review**<br>If child is incorrect, give opportunities to explore different fractional parts of identical whole objects, practically. |
| Q2 | Expressions and equations<br><br>MTH 1-15b | **On track**<br>318 g. Child shows understanding of missing addend problems and uses an appropriate strategy to solve this type of problem, for example by counting on or by using subtraction.<br><br>**Review**<br>Ensure child has a strategy to solve change unknown problems by giving examples involving smaller quantities that can be easily visualised. Ensure they understand = as balance rather than 'is the answer'. |
| Q3a | Time<br><br>MNU 1-10b | **On track**<br>Monday 7th October<br><br>**Review**<br>Maintain a visible calendar that the child can read daily. Give experience of looking ahead to identify dates. |
| Q3b | Time<br><br>MNU 1-10a | **On track**<br>Autumn<br><br>**Review**<br>Draw attention regularly to the changing seasons. Ensure the child knows order of seasons and months of the year. |

# End of first level assessments: answers and marking guidance

| Assessment | Topic | Question |
|---|---|---|
| Q4 | Ideas of chance and uncertainty<br><br>MNU 1-22a | **On track**<br>France because it is very far away, and it would be expensive. Accept any reasonable answer.<br><br>**Review**<br>Talk to the child to check their logic. Ensure they are exposed to the language of probability regularly in their everyday experiences. |
| Q5a | Data and analysis<br><br>MNU 1-20a | **On track**<br>Plays tig and plays football.<br><br>**Review**<br>Give child experience of Venn diagrams where there is data that belongs to more than one set, i.e. data in an intersection. |
| Q5b | Data and analysis<br><br>MNU 1-20c | **On track**<br>Draws Finlay outside both circles.<br><br>**Review**<br>Ensure child has experience of reading Venn diagrams that include data that belongs to neither set. |
| Q5c | Data and analysis<br><br>MNU 1-20c | **On track**<br>Draws Ailsa in left-hand section.<br><br>**Review**<br>Ensure child has experience of interpreting Venn diagrams and creating their own Venn diagrams. |